T0309496

Integrating
SCIENCE
With Mathematics
& Literacy

Second Edition

Elizabeth Hammerman • Diann Musial

Integrating
SCIENCE
With Mathematics
& Literacy

New Visions for
Learning and Assessment

**Second Edition of
Classroom 2061: Activity-Based Assessments in Science
Integrated With Mathematics and Language Arts**

Foreword by Robert E. Yager

CORWIN PRESS
A SAGE Company
Thousand Oaks, CA 91320

Copyright © 2008 by Corwin Press

For information:

 Corwin Press
A Sage Publications Company
2455 Teller Road
Thousand Oaks, California 91320
www.corwinpress.com

Sage Publications India Pvt. Ltd.
B 1/I 1 Mohan Cooperative
 Industrial Area
Mathura Road, New Delhi 110 044
India

Sage Publications Ltd.
1 Oliver's Yard
55 City Road
London EC1Y 1SP
United Kingdom

Sage Publications Asia-Pacific Pte. Ltd.
33 Pekin Street #02-01
Far East Square
Singapore 048763

Printed in the United States of America

Library of Congress Cataloging-in-Publication Data

Hammerman, Elizabeth L.
Integrating science with mathematics & literacy: new visions for learning and assessment/Elizabeth Hammerman, Diann Musial.—2nd ed.
    p. cm.
Rev. ed. of: Classroom 2061. 1995.
Includes bibliographical references and index.
ISBN 978-1-4129-5563-8 (cloth)
ISBN 978-1-4129-5564-5 (pbk.)
    1. Science—Study and teaching—Activity programs. 2. Science—Ability testing.
3. Mathematics—Study and teaching. 4. Language arts. I. Musial, Diann.
II. Hammerman, Elizabeth L. Classroom 2061. III. Title. IV. Title: Integrating science with mathematics and literacy.

Q181.H1473 2008
507.1—dc22   2007020450

This book is printed on acid-free paper.

07   08   09   10   11   10 9 8 7 6 5 4 3 2 1

Acquisitions Editor:	Hudson Perigo
Editorial Assistants:	Cassandra Harris, Jordan Barbakow
Production Editor:	Libby Larson
Copy Editor:	Teresa Herlinger
Typesetter:	C&M Digitals (P) Ltd.
Proofreader:	Dennis Webb
Indexer:	Rick Hurd
Cover Designer:	Lisa Riley

Contents

Foreword

New definitions and structures for science curriculum and new teaching strategies are difficult to attain. However, change in assessment strategies has been even more difficult. Teachers often return to quizzes, unit tests, and standard achievement measures as the only valid and objective indicators of learning. School administrators, boards of education, and the general public are also guilty of assuming that the only valid assessment strategies are the typical recall type of instruments and tests. New curriculum and teaching strategies are often assessed in terms of their effect upon standard measures of student learning.

For his doctoral dissertation at the University of Iowa in 1991, E. E. Zehr studied more than two hundred Iowa K–12 teachers who were actively involved with science-technology-society approaches in association with the National Science Teachers Association and the Science Standards and Curriculum project. He found that although teachers were willing to assess students in multiple domains, they used traditional test items to determine grades. Many attempted to write application and performance items; however, careful analysis revealed that nearly all of the assessment items resembled typical recall questions that focused on definitions, ideas, and skills taught directly in the classroom. These teachers seldom noted the importance of altering their assessment strategies.

Elizabeth Hammerman and Diann Musial recognize the problem with assessment reforms. They write and speak to the importance of assessment in meeting new goals, matching new curriculum structures, and conforming to the improvement in teaching practices. *Integrating Science With Mathematics & Literacy* provides a fine rationale for changing assessment to relate to the other facets of current reform efforts. The authors' philosophy, recommendations, and examples correspond to the ideas advanced in the National Research Council's *National Science Education Standards*.

Hammerman and Musial developed their book in collaboration with some of the most creative upper-elementary and middle school teachers in Illinois. The result is an easily understood and helpful book that encourages teachers to develop, try, and use performance measures in assessing their teaching and the learning of their students.

The authors define performance assessments as a set of tasks that include hands-on activities, criterion-referenced test items, and open-ended writing prompts. They are structured to match the type of inquiry exhibited by practicing scientists and mathematicians. Through the multidimensional tasks, students are given various opportunities to build concepts, practice and develop skills, and show what they know and can do.

The tasks both teach and assess content, skills, and habits of mind and expand students' abilities to relate science to the (real) natural world. The tasks teach because students are challenged to investigate concepts, collect data, make sense of these data, and apply their learning. Hence, even though these performance tasks are meant to provide assessment information, they are not boring or repetitive because as students demonstrate their knowledge, skills, and dispositions, they also acquire new knowledge and refine their skills.

It is important that students, parents, administrators, and others become involved in identifying desirable and appropriate learning and assessment performances. They should be partners in encouraging learning and determining when real learning has occurred. Real learning is usually best indicated by the transfer of concepts and skills to completely new situations.

Assessment is a basic part of the teaching/learning process. In fact, science cannot exist without it. Scientific inquiry begins with a question, followed by a personal explanation (hypothesis), which in turn must be tested for its validity. This testing is assessment; it is the evidence that must be shared and accepted by others for a hypothesis about the natural world to become scientific knowledge.

Engaging students' minds in the learning process is the goal. In the February 1994 issue of *Educational Leadership* ("How to Engage Students in Learning," pp. 11–13), Vito Perrone offers the following list of strategies for student engagement:

1. Students have time to wonder and find a particular direction that interests them.

2. Topics have a "strange" quality—something common seen in a new way, evoking a lingering question.

3. Teachers permit—even encourage—different forms of expression and respect students' views.

4. Teachers are passionate about their work. The richest activities are those "invented" by teachers and their students.

5. Students create original and public products; they gain some form of "expertness."

6. Students do something—e.g., participate in political action, write a letter to the editor, work with the homeless.

7. Students sense that the results of their work are not predetermined or fully predictable.

Hammerman and Musial provide ideas for performance assessments that promote student engagement. They offer great strategies for developing rubrics to determine the extent and the degree to which real learning has occurred.

I recommend *Integrating Science With Mathematics & Literacy* to all teachers who want to learn how to make their assessment of learning more authentic. The ideas will help promote current reform efforts and provide great hope that these efforts will be more successful and longer lasting than all our past attempts at educational reform.

Robert E. Yager, Professor
Science Education
The University of Iowa

Acknowledgments

We would like to acknowledge the following individuals who reviewed this book:

Charlotte Kenney
Eighth-Grade Math/Science Teacher
Browns River Middle School
Jericho, VT

Sheila Smith
Science Specialist/National Science Foundation Project Director
Jackson Public Schools
Jackson, MS

Melissa Miller
Science Teacher
Randall G. Lynch Middle School
Farmington, AR

Kathy Prophet
Science Teacher
Hellstern Middle School
Springdale, AR

Susan Chase-Foster
Seventh-Grade Language Arts Teacher/Professional Development Coach
Bellingham School District
Bellingham, WA

About the Authors

 **Elizabeth Hammerman, EdD,** through her love of science and her determination to raise the quality of standards-based science education for all children, has become a well-established science teacher educator, consultant, author, and workshop presenter.

In addition to her classroom teaching experience at the middle school, high school, and university levels, and her affiliation with numerous regional education centers and major organizations, she has offered undergraduate and graduate-level courses in science education and assessment for seven universities and worked extensively with K–12 teachers in their school systems to develop and implement standards-based instruction and assessment.

Recently, she wrote text, activities, and assessments for an online middle school science program that addresses all of the National Science Education Standards for life, earth and space, and physical sciences. She has given numerous presentations related to high-quality science teaching and learning at national and state educational conferences and has published articles dealing with assessment in science and successful science teaching in professional journals.

Elizabeth is committed to helping teachers offer high-quality instruction to enable students to attain higher levels of achievement. Her recent Corwin Press publications, *Eight Essentials of Inquiry-Based Science* (2005) and *Becoming a Better Science Teacher* (2006), were written to guide teachers through professional development initiatives leading to a better understanding of high-quality, standards-based science education.

 **Diann Musial** is a distinguished teaching professor of education at Northern Illinois University. She has an interdisciplinary educational history including a bachelor's degree in classical philosophy, a master's degree in physics, and a doctorate in education and has done post-doctoral work in social thought. Diann is an experienced mathematics and science educator and was an elementary school principal in Chicago, Illinois. She has directed over 20 professional development grants funded by Eisenhower, Scientific Literacy, and NCLB funds. Dr. Musial is currently working on a manuscript focused on the foundations of meaningful educational assessment.

*To Alton—our mentor who gave inspiration
and united two different worlds of thought*

*To Donald—our colleague who continually provides
guidance and support for our efforts*

*And with gratitude to Susan Klipp, a teacher who truly understands
how students think and learn, for her creative contributions*

Introduction

This book was written in response to many requests from schools to provide a clearly articulated set of performance tasks for teaching and assessment in science. Steeped in the sweeping ideas of Project 2061 as presented in the American Association for the Advancement of Science's *Science for All Americans* and *Benchmarks for Science Literacy* and the National Research Council's *National Science Education Standards*, we have formulated a set of carefully crafted prototypes to show the alignment of performance-based instruction with classroom assessment.

To provide opportunities for students to show what they know and don't know and what they can and cannot yet do throughout the instructional process, multiple and varied assessment methods must be employed. At the heart of these methods is the notion of a performance context. Through performance contexts, teachers can engage students in a series of multidimensional tasks that match the work of practicing scientists and mathematicians. By participating in such tasks, students develop new concepts or build on existing concepts and practice skills and thinking strategies within an inquiry context. Seamlessly, students demonstrate what they are learning through observable behaviors, notebooks or learning logs, drawings and illustrations, data analyses and explanations, peer interviews, teacher-student interviews, demonstrations, and projects. Educators are urged to employ a wide variety of assessment tools to create student profiles that are based on valid and reliable assessments.

We have designed a set of performance contexts that offer students the opportunity to demonstrate knowledge and multiple science abilities through investigation, data gathering and analysis, communication, application, problem solving, inventiveness, and persistence. The specific knowledge and science abilities are drawn from those identified in *Benchmarks for Science Literacy* (American Association for the Advancement of Science, 1993), *The National Science Education Standards* (National Research Council, 1996), and *Principles and Standards for School Mathematics* (National Council of Teachers of Mathematics Standards 2000 Project Writing Group, 2000). They are models for school districts and teachers to use in developing their own meaningful performance tasks for instruction and assessment. As prototypes, they illustrate a new way of looking at instructional activities and ongoing assessment in science, mathematics, and language. As assessments, they enable students to demonstrate their understanding of concepts, skills, and attitudes embedded within learning tasks.

The performance tasks in this book do not single out small individual components of the national standards; that is, they are not meant to measure one concept or skill at a time. Rather, these learning tasks are rich and filled with opportunities for students to develop and show their understanding of concepts, practice skills, and exhibit desirable habits of mind. The performances allow students to examine the world in a variety of ways while simultaneously yielding reliable performance data.

■ ASSESSING UNDERSTANDING

Humans strive to make sense of the world around them; it is part of human nature to seek understanding. Not surprisingly, both parents and students rely on the nation's schools to achieve understanding in the content areas. But what is understanding? This word is used in many different ways and contexts. If understanding is the desired outcome of instruction, educators must recognize the multidimensional aspects of the concept and carefully assess the congruence of their instructional practices to these different dimensions. For instance, in the context of science, knowledge may be viewed as having dimensions that relate to three levels of understanding: knowledge as information, knowledge as relationship, and knowledge as metacognition. An increasingly sophisticated level of knowing characterizes each dimension.

Knowledge as Information (Level One)

The first level of understanding includes a set of facts or information that one might accumulate and store internally as mental images. Informational knowledge is important merely as a first step toward reaching understanding. Traditional tests tend to focus on this level of knowledge for several reasons: (1) paper-and-pencil test items dealing with factual information are readily available; (2) testing time to engage in thought-provoking questions is not needed; (3) a broader range of topics can be assessed at the lower level; and (4) a set of facts is often all the teacher knows or has time to teach about a topic. More sophisticated levels of understanding are seldom assessed and, therefore, are left to chance.

Knowledge as Relationship (Level Two)

A more advanced level of knowledge goes beyond the information or content level that is part of one's mental image. This level focuses on the relationships between mental constructs and experiences. At this level of understanding, students are able to generate justifications and explanations for their mental images. For instance, when a student identifies rocks by their names and observable properties *and* knows the rocks in terms of their mineral compositions or the manner in which they were formed, the student has achieved a higher level of knowing. At this level, students are able to relate rocks to other curricular content and processes and to their formation in the natural world, thus demonstrating a broader and deeper understanding. Ultimately, the student views rocks as part of larger constructs such as matter, cycle, and change. This level of understanding can be effectively assessed in an activity-driven performance.

Knowledge as Metacognition (Level Three)

The third level of knowledge goes beyond what one already knows and the connections to related concepts and phenomena. It challenges the boundaries of the learner's mental images and experiences. Through inquiry, students are able to think about their own thinking, critique what they know, and generate new questions that reach beyond the knowledge they already possess. Such questions are generated by encountering new experiences in a somewhat "risky" manner, since challenging the limits of what one knows includes admitting that one does not have all the answers or knowledge. For example, in a physics class, students might study various types of forces and see them in action. Students may learn the concept of force through a knowledge base (Level One) and through firsthand experience at an amusement park (Level Two). The realization of force becomes a set of mental images coupled with experiential learning and problem solving. Once the student reaches a relational level of concept understanding within the realm of physics, the student can further investigate a concept through inquiry into the psychological and social forces operating within complex systems and in other contexts.

COMPONENTS OF MEANINGFUL
PERFORMANCE TASKS ■

The prototypical performance tasks in this book have been designed to allow students to move beyond the level of informational knowledge. Each task provides opportunities for students to explore their understanding of science, mathematics, and language through inquiry and applications. The performance tasks permit students to ask questions, investigate natural phenomena, and explore new relationships. The tasks enable students to exhibit at least two levels of understanding and allow teachers the opportunity to go beyond what is addressed in these activities. To accommodate important educational goals, each task has been carefully constructed to include the following components or criteria.

Thoughtful, Engaging Approaches

As far as possible, the performances promote higher-order thinking and help develop more complex cognitive functions. Students have an opportunity to display different levels of understanding during these performances, ranging from concept introduction to concept application. Students will ask questions, engage in investigations, exhibit understanding, make meaningful applications, and perform other tasks requiring complexity of thought.

Rich Opportunities for Problem Solving

The prototypical performance tasks allow students to solve problems and apply learning in a variety of ways. They encourage students to explore different problem-solving paths and demonstrate understanding in meaningful ways. As such, these performances include both guided inquiry and open inquiry tasks, active rather than passive learning experiences, and authentic rather than contrived contexts. In addition, they address essential rather than tangential learning goals.

Science Concepts

The performances are focused on the basic concepts of science at the fourth- through ninth-grade levels. The performance tasks address the unifying concepts and processes of science through the major topics that relate to them (e.g., magnetism, water cycle, transfer of energy, pollution, and the like).

Process Skills and Thinking Strategies

The performances are rich with the process skills of science, such as observing, classifying, measuring, making inferences, predicting, data collecting, and drawing conclusions, as well as complex thinking skills such as reasoning, comparing, and problem solving. The prototypical performances employ and assess a variety of skills within each task.

Habits of Mind

Throughout history, people have passed down values, attitudes, and perspectives about knowledge from one generation to another. These can be thought of as habits of mind because they relate to an outlook toward knowledge and learning and ways of thinking and acting. Habits of mind in the context of science include a willingness to take risks and the ability to formulate questions, exhibit curiosity, show persistence, analyze information with an open mind, and display a healthy skepticism. The prototypical performances enable students to develop and practice these habits of mind.

Science-Technology-Society Connections

The prototypical performances are set within a real-life context. Students are asked to investigate science problems and questions that are relevant to the society in which they live. The performances also require students to consider the relationships and applications of science principles to technology and to the local, state, national, and global society. In this work, the relationships will be referred to as S-T-S (science-technology-society).

■ THE PERFORMANCE TASKS IN THIS BOOK

Although the learning and assessment tasks in this book have strong science components, they have been constructed to reflect the new visions for mathematics and language literacy as well. The tasks include many of the components of mathematical power, as discussed in Chapter 2, along with the discourse that closely ties to language literacy. The analytic rubric provided for each performance task identifies indicators of learning that relate to the unifying concepts and processes of science, as well as indicators of learning from mathematics and language literacy. The rubric is described in detail in Chapter 3.

Chapter 4 is intended to guide you in understanding our prototypical models and to help you develop authentic performance tasks through which indicators of learning that are of interest and value to you can be assessed. This chapter also describes the importance of integrating instruction with ongoing formative assessment so that these tasks are seamlessly intertwined.

The performance tasks that comprise Chapters 5 through 13 provide a variety of ways to teach meaningful concepts and skills and assess students' understanding of concepts, acquisition of process skills, and ability to make real-world (S-T-S) applications. Each chapter includes one or more hands-on activities through which to develop concept understanding, skills, and applications. The tasks provide opportunities for learning and for ongoing assessment. For more formal assessment, a set of sample criterion-referenced test questions (most at the knowledge/ comprehension level) related to a few of the basic concepts and a writing prompt are offered. Suggestions are given for adding indicators of learning as needed and for scoring the writing prompts.

WHAT'S NEW IN THIS EDITION ■

The roles of standards and assessment in public school education have changed dramatically since this book was first written in the mid-90s. In this new and revised edition, the visions for science, mathematics, and language arts are more clearly articulated and aligned with national and state standards. The prototypic performance assessments have evolved into powerful models of standards-based instruction that engage students in meaningful learning and formative assessments that provide feedback to guide instruction.

The multidimensional performance tasks illustrate new ways of approaching instruction and assessment in science, mathematics, and language. As instructional activities, the tasks model inquiry-based science in ways that research has shown improves student achievement and they demonstrate a variety of ways to embed mathematics and literacy in science. As assessments, the tasks focus on important indicators of learning linked to standards and enable students to demonstrate their understanding of concepts, skills, and habits of mind within meaningful contexts.

PART I
Theory

What Is Science?

1

S cience is often thought of as a body of knowledge dealing with various aspects of the physical and biological world. Does this mean that the road to understanding each discipline of science is simply a journey through massive amounts of information and mastery of disconnected concepts?

To the contrary, more contemporary definitions of science and science education view concept acquisition as but one piece of a broader, richer picture. *Science for All Americans* (American Association for the Advancement of Science [AAAS], 1989) defines science education as education in natural and social sciences, mathematics, and technology. This larger vision of science focuses on inquiry about the physical world in the context of society. This vision can be thought of as a puzzle that comprises an intricate interplay of several pieces: concept understanding; process skills; habits of mind; and connections among science, technology, and society. The pieces of science, just as the pieces in a puzzle, are not meaningful in and of themselves. Rather, each piece must be studied within the framework of the other pieces.

SCIENTIFIC LITERACY ■

It is not enough for education to be self-serving. Science education needs to prepare citizens to deal with global, national, and local problems such as population growth; loss of resources; and the effects of pollution, disease, and social strife. The definition of science literacy includes the following characteristics (AAAS, 1989):

- Familiarization with the natural world
- Awareness of how mathematics, science, and technology are related to one another
- Understanding of key concepts and principles of science

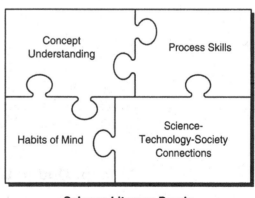

Science Literacy Puzzle

- Ability to think scientifically
- Knowledge that science, mathematics, and technology are human enterprises and what that implies about their strengths and limitations
- Ability to use scientific knowledge and ways of thinking for personal and social purposes

This hallmark description of science literacy, developed by the American Association for the Advancement of Science, was endorsed and enriched by the collaborative work of the National Research Council (NRC). In its 1996 release of the *National Science Education Standards* (NSES), the council stated that scientific literacy

> means that a person can ask, find, or determine answers to questions derived from curiosity about everyday experiences. This means that a person has the ability to describe, explain, and predict natural phenomena. Scientific literacy entails being able to read with understanding articles about science and to engage in social conversation about the validity of conclusions. Scientific literacy implies that person can identify scientific issues underlying national and local decisions and express positions that are scientifically and technologically informed. Scientific literacy also implies the capacity to pose and evaluate arguments based on evidence and to apply conclusions from such arguments appropriately. (p. 22)

△ *The key components of scientific literacy should be carefully and creatively integrated into the school curriculum.*

In order to accomplish the goal of scientific literacy, key components should be carefully and creatively integrated into the school curriculum. Such curriculum components derived from the list of characteristics above might include the following:

- A knowledge base emphasizing basic concepts and principles as well as more abstract concepts such as unity and diversity
- An understanding of the interrelationships of science, technology, and society (S-T-S connections)
- Strengths and limitations of science as a human enterprise
- The ability to think and trust the thinking process and to develop problem-solving skills and processes
- The ability to read and interpret science publications, evaluate arguments based on evidence, and apply conclusions

Take a moment to reflect on the science curriculum with which you work. Analyze it with regard to the components of scientific literacy noted above. The results of your analysis will assist you in determining the degree to which your educational program matches the new vision for science education. Such an analysis is a crucial step in the development of meaningful tasks for learning and assessment.

Concept Understanding

The traditional fields of science are biology, physics, chemistry, and earth science. Environmental science might be considered a separate category or it might be included in earth science or biology. While it might appear that these categories are universally understood since textbooks are generally organized under these headings, the arbitrary designations do little to define the concept knowledge base of science. For example, many branches of science, such as oceanography and ecology, have biological *and* physical aspects as interrelated components of study.

A more appropriate portrayal of the concept areas of science is as a set of universally accepted unifying concepts and processes or "big ideas" that thread throughout major areas of study of the natural world. When viewed in this larger, broader context, a set of cognitive "frames" is identified that becomes the foundation for building and structuring knowledge over time (Musial & Hammerman, 1993).

Science curriculum is often presented in the form of topics, but most commonly taught topics are merely subcategories of bigger ideas, also called unifying concepts. An example of a set of related unifying concepts is "system, order, and organization." An important concept underlying these unifying concepts is the notion of "cycle." A cycle can be defined as a continuous series of events that repeats itself over time. The events in a cycle can be repeated at regular or variable intervals. Thus, a cycle is a type of system having order and organization. At the primary school level, the cycle of day and night may be introduced. The regularity of the day and night cycle can be easily observed in a short period of time. The cycle of the seasons takes longer to observe, but over the course of the year students can recognize the varying characteristics of seasonal change. Some cycles, like the water cycle, do not occur at regular intervals but there is, nevertheless, order and organization. Some life cycles can be observed in a relatively short time span, such as the life cycle of the darkling beetle, but other cycles, such as the rock-soil cycle or certain chemical cycles, are not as regular or as easily observed firsthand.

Regardless of which examples of cycles are studied, students need to understand that all the examples represent the larger unifying concept of system (with order and organization). The unifying concepts are also called cognitive frames of thought such as those that may be introduced early in a child's life and known through just a few examples. Over time, cognitive frames expand and develop through exposure to multiple examples, experiences, and connections. Every cognitive frame has a broad meaning that students will use to make sense of a more abstract, complex world.

The characteristics common to the cycle of the seasons and the life cycle of the frog may not be apparent to most students. However, an effective, integrated curriculum will identify similarities and differences between the examples and recognize their relationships to the unifying concepts. Unifying concepts are useful because they go across disciplines. Examples of cycles are found in history, mathematics, economics, and other areas of study.

Existing commonalities between the knowledge base of science and other disciplines provide a natural way to classify information so that the brain can recall it more easily. Information relating to

> Δ A more appropriate portrayal of the content areas of science is as a set of universally accepted unifying concepts and processes or "big ideas" that thread throughout major areas of study of the natural world.

Unifying Concepts and Processes of Science:

- Systems, order, and organization
- Evidence, models, and explanations
- Constancy, change, and measurement
- Evolution and equilibrium
- Form and function

science becomes far more manageable when students realize that most of it will fit into a set of unifying concepts and processes.

Process Skills and Thinking Strategies

The process skills of science are ways of thinking and acting used by scientists in their work. On a basic level, observation, classification, and prediction skills are used. At a higher level, the thinking skills of hypothesizing, controlling experimental variables, and drawing conclusions engage students in learning about the natural world. The inquiry approach to learning science is rich with process and thinking skills that allow students to become active learners, make use of their senses, and become involved emotionally. Thus, students are able to make more meaningful connections.

Process and thinking skills are embedded in the procedures of an activity. The names of the skills should be used and reinforced by teachers so that each skill name becomes familiar to students, enabling them to use the vocabulary when they explain their learning. Although there is no universal list of scientific process skills or complex thinking skills, Figures 1.1 and 1.2 provide samples of those found in the literature.

Science Process Skills

Observation: the use of one or more of the senses to identify properties of objects and natural phenomena

Classification: a system or method for arranging or distributing objects, events, or information

Making Inferences: giving explanations for an observation or conclusions based on logic and reasoning

Prediction: forecasting future events or conditions

Measurement: making quantitative observations by comparing an object, event, or other phenomenon to a conventional or nonconventional standard

Using Numbers: counting and creating categories; applying mathematical rules or formulae to quantities

Creating Models: using two- or three-dimensional graphic illustrations or other multisensory representations to communicate ideas or concepts

Defining Operationally: naming or defining objects, events, or phenomena on the basis of their functions and/or identified characteristics

Identifying Variables: recognizing factors or events that are likely to change under certain conditions

Formulating Hypotheses: making statements that are tentative and testable; a special type of prediction that suggests relationships between variables

Recording and Interpreting Data: collecting, storing (through writing, drawing, audio or visual display, etc.), and analyzing information that has been obtained through the senses

Drawing Conclusions: making summary statements that follow logically from data collected throughout an experience or experiences

---

Figure 1.1

Complex Thinking Skills

Comparing: describing objects and events by their properties and determining how things are alike and how they are different (Venn diagrams are often used to show comparisons.)

Creating Representations: generalizing a pattern of information and representing it in another way such as through the use of graphic organizers

Making Analogies: designing and describing ways that objects and events are alike (categorically or otherwise) to illustrate an understanding

Reasoning: drawing inferences or conclusions from known or assumed facts

Problem Solving: designing and explaining a possible solution or solutions when given a set of conditions or circumstances (Technological design and Science Olympiad activities engage students in problem solving.)

Inventing: designing a product or process that shows a deeper understanding or applying a concept

Meta-Cognition: reflecting on one's own thinking. Carefully crafted units of instruction include questions for discussion, reflection, and interpreting meaning.

Figure 1.2

Habits of Mind

Another important component of science is a set of attitudes and dispositions referred to as "habits of mind," which relate to how a student views knowledge and learning. Scientists learn about the universe through direct involvement in laboratory and field investigations. The behaviors, attitudes, beliefs, and ways of thinking exemplified by the working scientist can be practiced by students through inquiry-based science instruction. Dispositions such as curiosity, honesty, integrity, open-mindedness, respect for life, willingness to suspend judgment, and respect for data are some of the habits of mind students can develop through an inquiry-based program. Students can also practice safety, accuracy, good experimental technique, systematization of data, persistence, effective communication, analysis of strategies and results, and replication of the work of others. When students are allowed to work like scientists, they have the best chance of developing habits of mind that are highly regarded by those in science and by society.

Science-Technology-Society Connections

Understanding the relationship of science to technology and society is also a critical component of scientific literacy. Science does not exist in a vacuum. Science investigates the world around us; it exists everywhere. Interest in science seems to wane when it is not seen as a vital component of our lives. As humans, we are a part of the natural world; to understand nature is to better understand ourselves.

△ **When students recognize these science-technology-society connections (S-T-S), they begin to appreciate the value of scientific discoveries and become aware of the strengths and limitations of science in relation to social problems.**

In our current technological world, technology and society are intertwined with science in ways that are both intriguing and masterful. Yet, students are not always aware of how the concepts of science apply to their lives and to society. They do not fully understand the ways that politics, values, economics, and other social structures interact with the findings of scientists. When students recognize these science-technology-society connections (S-T-S), they begin to appreciate the value of scientific discoveries and become aware of the strengths and limitations of science in relation to social problems.

Most students are aware of some of the areas of science that operate in the nation or world, such as the exploration of outer space or the hunt for the fossils of extinct dinosaurs. Through the media, students are exposed to scientific fiction as well as fact. It is critical that students understand the differences between science fact and science fiction. When students study significant scientific theories, science history, and the evolution of science, they recognize the value of obtaining and relying on evidence.

■ METHODS OF TEACHING SCIENCE

One major factor that influences a student's interest in and respect for science is the way that the student is taught. Traditional science instruction often consists of reading, writing, and memorizing facts. In contrast, inquiry-based science provides students with exciting opportunities to explore, investigate, research, and discover, which are natural to the learning process. The goals of science education can be met through a strong commitment to an inquiry-based science program as the means for learning standards-based concepts, developing skills, practicing scientific habits of mind, and making real-world connections.

New Visions for Mathematics and Language Literacy 2

A revolution is taking place within the fields of mathematics and language literacy. Both have experienced a paradigm shift. This shift includes a move away from the mastery of information to the attainment of a broad, interconnected knowledge base. Both fields are being restructured toward a new vision of knowledge in which the emphasis is placed on providing instruction that enables the learner to analyze and communicate effectively in an ever-changing environment. To create authentic assessments, the new visions for mathematics and language literacy must be carefully examined and effectively applied.

THE VISION FOR MATHEMATICS ■

Perhaps more than any other discipline, mathematics has evolved into a massive information base filled with centuries of insights. Somewhere along its development, mathematics began to look like an encyclopedia of information about relationships among numbers and space. Schools began to amass this information into thicker and thicker textbooks filled with problems that required one right answer. Not surprisingly, teachers began to teach for these right answers and students began to view mathematics as the mastery of problems that someone else developed. The problems became more and more abstract, and their relationship to the real world became increasingly remote.

Recent efforts by the National Council of Teachers of Mathematics (NCTM) and the National Research Council (NRC) have led mathematicians, educators, and the general public to reconsider the essence and vision of mathematics. This effort has resulted in the development of two terms: *mathematical literacy* and *mathematical power*. Mathematical literacy includes having an appreciation of the value and beauty of mathematics and being able and inclined to appraise and use quantitative information. Mathematical power is the ability to do purposeful and worthwhile work, such as exploring, conjecturing, and using logical reasoning. It is also the ability to use a variety of mathematical methods effectively to solve non-routine problems and to possess the self-confidence and disposition to do so (NCTM Commission, 1991).

Mathematical Literacy and Mathematical Power

As suggested above, the new vision for mathematics revolves around the terms *mathematical literacy* and *mathematical power*. This vision places mathematical reasoning, problem solving, communication, and connections to the real world at the center of mathematics. Convergent thinking is de-emphasized and replaced with divergent thinking. Mathematical problems with single answers are replaced with interdisciplinary problems requiring several thinking paths that lead to a variety of reasonable solutions. Computational algorithms, the manipulation of expressions, and paper-and-pencil drills must no longer dominate mathematics instruction. Beyond the standard fare of number concepts and operations, the school curriculum must include serious exploration of geometry, measurement, statistics, probability, algebra, and functions. Students should encounter, develop, and use mathematical ideas and skills in the context of genuine problems and situations. In doing so, they will develop the ability to use a variety of resources and tools, such as calculators; computers; and concrete, pictorial, and metaphorical models. Students must know and be able to choose appropriate methods of computation, including estimation, mental calculation, and various types of technology. As they explore and solve problems, students should be able to engage in conjecture and argument.

The new vision for mathematics is similar to the new vision for science. As these two disciplines attempt to recover their essences in the midst of a sophisticated technological society, they begin to focus on the similar components of concept understanding, thinking or process skills, and dispositions or habits of mind. An examination of these components reveals the remarkable relationship between science and mathematics. Assessments in science that focus on these basic components will be potential assessments in mathematics as well.

Δ The new vision for mathematics is similar to the new vision for science. As these two disciplines attempt to recover their essences in the midst of a sophisticated techno-logical society, they begin to focus on the similar components of concept understand-ing, thinking or process skills, and dis-positions or habits of mind.

Mathematics Understanding—Big Ideas

The first component of the new vision for mathematics relates to concept understanding, where the concepts are composed of big ideas. This component is labeled content standards, and it calls on teachers to focus on five large mathematical ideas:

- Numbers and operations
- Algebra
- Geometry
- Measurement
- Data analysis and probability

Each of these content standards applies to all grades (prekindergarten through Grade 12). These large ideas or content standards are not to be taught in isolation; rather, the NCTM standards emphasize that students should use these mathematical concepts to make connections among other concepts and across disciplines. The intention of this set of standards is to encourage teachers to focus consistently on these big ideas so that growth in students' knowledge and sophistication occurs consistently as they progress through the curriculum. The large ideas abound with smaller concepts, and these concepts are taught under the framework of the larger standards. Supporting concepts such as change, symmetry, equality, inequality, numeration, ratio, perimeter, and area are presented as important dimensions or pieces of these larger content standards.

Mathematical Thinking or Process Standards

The second component of the new vision for mathematics emphasizes thinking or process skill development. These standards are presented as ways of acquiring and using content knowledge and include the following:

- Problem solving
- Reasoning and proof
- Communication
- Connections
- Representation

The process standards focus on the use of knowledge and understanding to analyze, conjecture, design, evaluate, generalize, investigate, model, predict, transform, or verify. This component of mathematical power is almost identical to that of science process skill development. In fact, the process skills for science can be found in the list of mathematical skills generated by both the National Council of Teachers of Mathematics and the National Research Council.

This emphasis on the development of thinking skills has significant implications for assessment. Teachers cannot assess thinking strategies without providing problems that require the use of these skills. Although memorization is a useful skill, it is not the focus of the new vision for mathematics. Performance assessments that require skills such as analysis, conjecture, speculation, and reflection will evoke student thinking more effectively than assessments composed of paper-and-pencil exercises.

Mathematics Tools and Techniques

Given the technological complexity of today's real-life problems, the new vision for mathematics includes an emphasis on efficiently and effectively solving problems.

This vision requires teachers to assist students in the use of manipulatives. Using calculators and computers is no longer a tangential component of mathematics; it is an essential part of a student's mathematics education. Problems that students encounter in the classroom should be complex, but students should no longer be rewarded for solving such complexities the "long way." Instead, students should be encouraged to consider which tools can help to solve problems most efficiently in a given situation.

△ *Using calculators and computers is no longer a tangential component of mathematics; it is an essential part of a student's mathematics education.*

The new vision for mathematics also encompasses the use of diagrams, tables, charts, and other concrete materials to assist in solving problems. The use of number symbols is not always the most effective way to present a solution. Graphs, drawings, and written or verbal presentations are often more effective than numeric expressions.

Such a changed vision has profound implications for assessment. Performance assessments need to focus on problems that require the collection of data and the use of organizing structures such as charts, diagrams, and tables. Manipulatives should be available to students as needed but should not be used indiscriminately. Students should be encouraged to select the correct tool(s) based on the demands of the problem. To use the calculator for every problem is inappropriate, as is solving all problems through paper-and-pencil or mental computation.

Communication Skills

The final component of the new vision for mathematics focuses on communication. This component requires students to communicate results with various audiences for a variety of purposes. To fully assess mathematics without also assessing language skills is virtually impossible. As a result, the new vision for mathematics requires interdisciplinary assessments. To adequately assess learning in mathematics, teachers must assess language (specifically communication skills) simultaneously.

■ THE VISION FOR LANGUAGE LITERACY

Language literacy, like mathematics, has suffered because of its highly developed knowledge base. Throughout the last century, language literacy instruction has grown to include huge amounts of information about decoding, phonics, vocabulary, reading, sentence structure, writing, literature, poetry, and more. Over time, language literacy has been specialized and compartmentalized into smaller subdisciplines. Reading has been separated from writing and literature from the development of language. However, efforts by the International Reading Association (IRA) and the National Council of Teachers of English (NCTE) have bolstered support for the "whole language" movement. As the word "whole" implies, this movement encourages teachers and students to view language as an interrelated set of activities that focuses on meaning-making and effective communication. Within this vision, reading cannot be taught separately from writing, nor can language be taught separately from literature. The whole-language movement has transformed the way textbooks are written and the way literature is presented.

The whole-language method encourages teachers to approach language literacy instruction as a threefold process: meaning-making, comprehending the meaning of others, and communicating meaning to others. This process requires students to actively reflect on the meanings they are developing. Students who are taught to reflect on the meanings that they create tend to develop meanings that are more accurate, comprehensible, and useful to society. Language literacy instruction assists students in this reflective process by providing a variety of symbols for expressing and clarifying meaning. For example, students can draw, create graphs, sing

△ *Students who are taught to reflect on the meanings that they create tend to develop meanings that are more accurate, comprehensible, and useful to society.*

songs, write poetry, and even use their bodies to express concepts that sometimes cannot be expressed in narrative form.

Language literacy also encourages students to actively develop their ability to understand the meanings of others. Such understanding will occur if students are taught to use and identify many different types of symbols. Sensitivity to different ways of communicating helps students comprehend the meanings of others. Assessments that are true to this vision of language literacy are interdisciplinary. Each time a student is asked to communicate, no matter what the discipline, that act of communication is part of the whole language. Mathematics, science, and language literacy are inextricably connected in the world of whole language.

Within this framework of viewing language literacy as a powerful tool for representing our ideas to the world and the world to us, the International Reading Association and the National Council of Teachers of English (1996) have released an interactive, integrated model for language literacy. In this model, they posit three dimensions that interact with each other in the development of language literacy: content, purpose, and development.

The content dimension addresses what a student should know and be able to do with the English language arts. This dimension includes knowledge of written, spoken, and visual texts and of the processes involved in creating, interpreting, and critiquing such texts.

The purpose dimension addresses the question of why we use language. In other words, it considers the range of motives, reasons, and desired outcomes to which we direct our literacy practices. We all use language for a variety of purposes, such as to learn, to express ideas, to convey information, and to persuade others. Any literacy event may involve several of these purposes.

The development dimension focuses on how learners develop competencies along the way. Students grow as language users by building knowledge of content and a repertoire of strategies such as predicting, synthesizing, reflecting, and identifying words and their meanings.

This interactive model for language literacy has compelling implications for the context in which language should develop. Clearly, a real-world context provides an excellent way to help students develop language literacy. This is so because in a real-world context, multiple opportunities surface for language to be used for different purposes. In fact, the IRA and NCTE (1996) language standards affirm the importance of "authentic learning experiences that involve a variety of contexts" (p. 14). The standards call on educators and policy makers to articulate curricula that generate learning opportunities that respond to local needs and interests.

IMPLICATIONS FOR TEACHING AND ASSESSMENT IN MATHEMATICS AND LANGUAGE LITERACY

To accommodate the new visions for mathematics and language literacy, assessments must be reconceptualized. Mathematics and language literacy assessments often focus on mechanics using a set of paper-and-pencil problems. To reflect a reformed vision for these fields, assessments need to be reformulated into significant tasks. These tasks provide the intellectual contexts for students' mathematical, scientific, social, and language development. They need to provide the stimulus for students to think in real-world contexts. Tasks that require students to reason and to communicate mathematically, scientifically, and socially are more likely to improve students' ability to make connections and increase their understanding. Teachers need to select activities that create opportunities for students to develop cognitive knowledge, skill competence, increased interest, and positive dispositions. Activities conducted in science classes can also focus on mathematics and language literacy, thus providing greater opportunities for developing interdisciplinary understandings.

Teaching and assessment tasks that capture the new visions for mathematics and language literacy need to be open-ended enough to permit discourse. Discourse refers to the multiple ways of representing, thinking, talking, agreeing, and disagreeing that teachers and students use to engage in the tasks. Discourse permits thinking to unfold and also affirms the importance of interactive, collaborative learning. Brainstorming, discussing, sharing results, and respectful questioning are skills that need to be practiced and sharpened. These skills relate to the communication aspect of mathematics and science; they also permit teachers to assess student understanding. Assessments rooted in activities that permit discourse will more readily lend themselves to the new visions for mathematics and language literacy.

Developing Rubrics for Teaching and Assessing Performance Tasks

3

As shown in the previous chapters, the boundaries that have traditionally separated the disciplines of science, mathematics, and language are being challenged. Science now focuses on a set of unifying concepts and processes, as does mathematics. The ideas look different in the contexts of the different disciplines, but they are essentially the same as they view topics in terms of the larger constructs of systems, order and organization, models and explanation, form and function, and so on. The new visions require teachers to reconsider the concepts they currently teach and reframe them in light of unifying concepts, or big ideas, that transfer across disciplines.

The same is true for process and thinking skills. It is now apparent that thinking skills required in science are similar to those needed in mathematics and other areas of the curriculum. For example, classification in mathematics is similar to classification in science. Analyzing a problem in mathematics is not unlike analyzing an investigation in science.

This is also true for the scientific habits of mind. In mathematics, these habits of mind are called mathematical dispositions. No matter what the term, these values and attitudes tend to be similar throughout the disciplines. For example, the ability to ask clear questions is essential in mathematics, science, and language arts. Noting similarities among the concepts, skills, and habits of mind of the different disciplines is helpful when developing rubrics. It will assist the teacher in identifying the important features of an interdisciplinary task.

■ RUBRICS

One of the greatest challenges in assessing learning through performance tasks is to identify the critical knowledge and skill features of the task. This can be likened to taking an x-ray. An x-ray enables a physician to view the underlying structures that support the human body. In a sense, this is the view needed when determining how to assess a performance. Every performance task contains important underlying features or indicators of the critical dimensions of learning and inquiry. In science, these dimensions include concept understanding, process skills, habits of mind, and science-technology-society connections. Teachers must step back from the performance and, like an x-ray, look beneath the surface to focus on the important dimensions. Identifying these indicators is at the heart of conceptualizing rubrics.

Rubrics can be thought of as a set of criteria that provides direction in determining what students know and are able to do. In their simplest form, rubrics are nothing more than an answer key for a multiple-choice test. In this case, the only rule is to count the number of correct answers and perhaps cluster the answers into different subtest scores. The rubrics for multiple-choice, paper-and-pencil tests are simple. However, developing rubrics for a rich performance task requires more thought and consideration. Performances can be assessed with holistic, generalized, or analytic rubrics.

Holistic Rubrics

A holistic rubric requires the teacher to think about or consider an entire task as a single entity or construct. Some holistic scoring experts maintain that holistic rubrics should not be developed prior to administering the performance task; the focus should be on the performance, not on predesigned rubrics. These experts encourage teachers to spend more effort developing rich performance opportunities for students than designing tasks to match predesigned rubrics.

To develop holistic rubrics, use the following steps:

1. Identify a worthy task.

2. Have students perform the task.

3. View performances of students in their entirety.

4. Separate performances into two groups: those that were adequate and those that were not.

5. View the adequate performances and select subgroups that stood out as exceptional and those that were merely acceptable.

6. Review the inadequate performances and separate them into two subgroups: those performances that display serious problems and those that are simply inadequate.

7. Return to each of the four groups, ranging from seriously inadequate to exceptional, and identify (based on the evidence of student work) the characteristics that best describe each category or group. These descriptions constitute the final rubric used to score other student work on the

same performance. Performances that represent the salient characteristics of each of the four groups can be selected and used to better clarify the different categories of the rubric.

Generalized Rubrics

Some experts contend that the disciplines are innately interconnected through similar dimensions and, therefore, generalized statements can be written that relate to a variety of performances within the same discipline or across disciplines. They contend that generalized rubrics used to assess concept understanding, process skills, habits of mind, and S-T-S connections are best designed by viewing the student's entire performance. These experts argue that when teachers break up a performance into discrete pieces, or indicators, they may be distracted from appreciating the beauty and continuity of the entire task. Generalized scoring rubrics provide a framework for assessing a task without explicitly identifying discrete indicators.

A generalized scoring rubric requires the teacher to consider the dimensions of a task as a single entity. Figure 3.1 identifies the dimensions of a performance task in terms of (1) the understanding of concepts and principles including the use of terminology, the quality and accuracy of explanations, and the quality and accuracy of representation and (2) the skills of inquiry including the use of processes and strategies, the use of tools and technology, and the ability to make connections. The rubric consists of categorical descriptions on a continuum. Teachers select the category along the continuum that best represents the dimension as exhibited in the performance.

Analytic Rubrics

The identification of individual knowledge and skills features critical to and inherent in a task allows teachers to assess concept understanding, process skills, and habits of mind as separate components. This discrete approach to developing rubrics is called the analytic method. For any task, the teacher decides which of the many features within the task will be assessed. Using the dimensions of concept understanding, concept application, thinking strategies and process skills, habits of mind, and S-T-S connections as a guide, the teacher lists student behaviors or indicators of learning that relate to any of these important learning dimensions. For each indicator, the teacher assigns a score on a scale such as "complete," "almost," and "not yet." "Complete" means that the student exhibited the indicator, "almost" means that there is some evidence that the student exhibited the indicator but something is incorrect or missing, and "not yet" means that the student did not show evidence of learning for that indicator. Indicators of learning that might be assessed in a performance task include the following:

- Making reasonable predictions
- Making and recording observations
- Using measurement equipment properly to obtain accurate data
- Collecting and recording data
- Analyzing data and drawing conclusions
- Making graphs or creating representations to show relationships
- Applying concepts to new situations

Figure 3.1 A Generalized Rubric for Performance Tasks in Science

Indicators of Learning	Novice	Apprentice	Practitioner	Expert
Understanding of concepts/principles				
Use of Terminology	~little/no use of terminology ~inappropriate use to describe &/or explain	~some use of terminology ~some used inappropriately	~uses available terminology ~most used appropriately	~uses all available terminology ~all used appropriately
Explanation (communication also assessed in inquiry section)	~explanation or conclusion faulty or inappropriate	~accurate explanation or conclusion with no support	~accurate/clear explanation or conclusion with some support	~accurate/clear explanation or conclusion with excellent support &/or elaboration
Representation	~little/no use of graphic organizer(s) to organize data &/or show thought patterns or inappropriate use of graphic organizer(s)	~some attempt to use graphic organizer to show data &/or thought ~obvious errors	~good use of graphic organizer(s) to show data &/or thought patterns ~all/most used appropriately ~some errors	~excellent use of graphic organizer(s) to show data &/or thought patterns ~all graphics labeled & used appropriately
Inquiry*				
Use of Skills and Strategies (includes ability to understand & do inquiry and to apply & communicate findings)	~shows little/no evidence of use of skills &/or strategies or ~inappropriate use of skills &/or strategies	~shows some use of skills &/or strategies ~some used appropriately	~shows good use of skills &/or strategies; most used appropriately	~shows excellent & appropriate use of all skills &/or strategies
Use of Tools and Technologies	~shows little/no or inappropriate use of tools or technologies (rulers, balances, magnifiers, A/V, models, etc.)	~shows some use of tools & technologies ~some used appropriately	~shows good use of tools & technologies ~all/most used appropriately	~shows excellent & appropriate use of all tools & technologies
Making Connections	~little/no connections made to technology &/or society ~inappropriate examples given	~some connections made to technology &/or society ~some/most examples are appropriate	~good connections of science to technology & society ~all/most examples appropriate	~excellent connections of science to technology & society ~great examples ~ability to elaborate

*Inquiry: the ability to use process skills and thinking strategies and to use tools and technologies in a variety of contexts

Analytic scoring is simple and efficient. Such an approach not only clearly identifies the different dimensions of a task (understanding related to the unifying concepts, process and thinking skills, habits of mind, and S-T-S connections), but may also identify the specific features or indicators (measures, observes, predicts, describes, analyzes, relates, works collaboratively, and so forth) for each of the dimensions that the teacher wishes to assess through the performance. These indicators can provide insight into a student's strengths and weaknesses and help diagnose problems related to a student's concept understanding, skills, or behavior. For example, for a performance task in which students are asked to identify a relationship between two variables, such as the effect of temperature on plant growth, the teacher could use the scoring scale mentioned earlier: "complete," "almost," or "not yet." The scale would provide immediate feedback to students, enabling them to reconsider their data analysis, continue with their project, or replicate their experiment.

Some complex indicators of learning such as "the student conducts a controlled experiment," involve multiple tasks and complex reasoning. In such cases, it may be more appropriate to show subcategories of specific indicators using bullet points, such as the following: identified the problem, made a hypothesis, identified and controlled variables, used appropriate materials, conducted multiple trials, made precise measurements, created an appropriate data table, labeled the table correctly, and so forth. This approach permits the specific indicators or behaviors to be listed under the larger dimensions of learning. Most important, the rubric should provide ample information to the students to both guide their work and provide specific feedback concerning the quantity and quality of their work.

RUBRICS FOR THE PERFORMANCE ASSESSMENTS IN THIS BOOK ■

We have determined that analytic rubrics are most useful for clarifying the components of performances. Analytic rubrics are user-friendly and enhance reliability between raters. For each performance in this book, we have identified the indicators that relate to the unifying concepts and processes as well as other science and mathematics learning dimensions and criteria for the writing prompts.

As part of a formative assessment, we have also provided sample criterion-referenced test items for each of the performance tasks. These tests are carefully structured to provide two to three questions that measure a single concept. In a criterion-referenced test, students are given multiple opportunities to show that they know a specific concept. We recommend that when using these tests, you be sure to carefully examine the answers to the three questions that assess the same concept. Only if students answer *all* the questions correctly do you have sufficient evidence that they know that particular concept. If the students get two of the questions about a concept correct, you might also infer that they understand that concept, but it may be wise to note the reason for the error in the third item. Hence, for a test with nine questions, there are really only three (not nine) concepts that are assessed. You may use the test as presented, or you may prefer to customize the test based on your needs. The answers are provided in Appendix A.

The important point to note about rubrics is that they identify the types of learning that can be observed through the task. Although the performance tasks provide multiple opportunities to observe different dimensions of knowledge and indicators of learning, only those that are related to the educational goals for the instructional unit need be included on the rubric. Teachers do not need to take note of every behavior or indicator of learning for every task. Only the individual teacher can determine which of the many learning opportunities inherent in the tasks are to be included and assessed through the rubric.

Putting It All Together 4

*Integrated Performance Tasks
for Learning and Assessment*

The tasks of teaching and assessing student learning have often been relegated to having students read from textbooks and recall information on paper-and-pencil tests. End-of-chapter tests are composed of questions that can be answered by memorizing terms or passages from the text. Even when laboratory activities are included in the instructional process, the measure of student achievement is often tied to recalling factual information, identifying a correct response to a question, restating the proper definition, or applying a mathematical formula correctly. Assessment is generally measured in terms of what the student is able to show at the knowledge/information level, and in science the acquisition of facts determines how informed students are.

A CALL FOR REFORM ■

Documents such as *Science for All Americans* (AAAS, 1989) call for reform in science education, emphasizing the relationship of science to mathematics and technology and recognizing the importance of the knowledge, skills, and habits of mind associated with all of these disciplines. National and state leaders emphasize the need to view science as more than a set of accumulated facts and theories. This recognition of science as multidimensional with connections to mathematics and technology requires administrators and teachers to redefine learning goals, revitalize instruction, and redesign assessments in ways that go beyond traditional forced-choice or short-answer tests.

The performance tasks in this book provide examples of a new type of classroom assessment that does more than assess. We call them "performance tasks for learning and assessment" because they both teach and assess in a seamless manner. Performance tasks in this new mode are no longer limited to end-of-year or end-of-chapter paper-and-pencil tests, nor are these tasks limited to measuring the learning of memorized material. Traditional classroom assessments often fail to provide the opportunity for students to show their broad range of concept understanding, skill acquisition, habits of mind, and connections to other disciplines and society at large. Meaningful assessments should be interesting and engaging in order to elicit information about what students know and can do as they are learning. This allows teachers to formatively make changes to their instruction in an effort to meet students' learning needs.

Grant Wiggins (1998) spoke of this need to develop a different type of meaningful, formative assessment in his call for learner-centered assessment. Wiggins states that in order for students to achieve better performance, educators need to teach in such a way that students can systematically self-correct their performances. He labels this type of learning and ongoing self-correcting behavior as *educative assessment*. Such tasks are to have two essential qualities:

1. They are anchored in *authentic performances*, namely tasks that teach students how adults are actually challenged in the field.

2. They provide students and teachers with feedback and opportunities that they can readily use to revise their performance on these and similar tasks.

In this text, we present specific examples of learner-centered performance assessments. Our tasks are authentic in that they are designed to represent the processes and habits of mind that working scientists and mathematicians use. Students are challenged to ask questions, solve problems, collect data, draw conclusions, apply knowledge, communicate results, and use mathematical and technological tools. Our tasks both teach and assess; they provide feedback along the way to the teacher and to the student. The feedback to the teacher allows him or her to monitor and guide the learning process and give timely feedback to students as they are learning new knowledge and new skills.

■ ASSESSMENT STRATEGIES THAT PROVIDE FEEDBACK TO TEACHERS AND STUDENTS

In addition to the call by Grant Wiggins to transform the meaning of assessment from a final evaluation into student-centered, self-reflective assessment, there is an equally strong call from other writers for classroom assessments that are formative in nature. Such assessments integrate teaching, learning, and assessment into a single event. Formative assessments are sprinkled throughout the act of teaching and learning and are intended to uncover students' learning difficulties so that teachers can adapt their own work to meet students' needs. In a carefully developed meta-analysis, Paul Black and Dylan Wiliam (1998) exposed the power of such formative assessments. Policy makers, teachers, and administrators were called upon to radically reconsider the types of assessments that are used in today's classrooms. Black and Wiliam claim that formative assessments must be varied and carefully tied to every teaching and learning act. Further, they challenge teachers to provide ongoing feedback so that students can self-correct as they learn.

Tasks that include a variety of feedback strategies provide more opportunities for student learning. The more varied and multidimensional the opportunities for students to learn and to communicate understanding, the more feedback can be offered. In this instance, more is better. Rich assessment plans might include the following strategies:

Learning logs or notebooks—student-generated descriptions, drawings, data, charts, inferences, generalizations, explanations, graphic organizers, other notes, research, and visuals that are documented during the learning process

Portfolios—a compilation of students' best work over time based on a variety of criteria

Interviews—face-to-face discussions (student-to-student or student-to-teacher) about student learning

Observations—recorded evidence of student learning or behaviors exhibited through classroom activities

Criterion-referenced tests—multiple-choice questions clustered according to selected criteria allowing students several opportunities to display their knowledge about each criterion

Writing prompts—authentic situations or questions that require students to apply their understanding, link learning to technology and to society, and communicate effectively through writing

Products—integrated, student-designed works that represent and summarize student understandings of concepts

Data collection and interpretation activities—experiences that require students to manipulate materials and reflect on their processes and data in meaningful ways

Our performance assessments are multifaceted tasks that focus on important dimensions of learning in a more comprehensive manner than any of the individual assessment strategies cited above. The performance tasks in this book comprise inquiry and learning activities that are interspersed with a variety of assessment strategies and combined into a meaningful whole. Such performance learning assessments provide a rich profile of student learning and supply valuable formative assessment information that both teachers and students can meaningfully use to enhance learning.

DESIGNING INTEGRATED PERFORMANCE TASKS FOR LEARNING AND ASSESSMENT

Meaningful performance assessments allow teachers to assess concept understanding, process skills, and habits of mind through a cohesive set of activities. These learning assessments have the same qualities and characteristics as good instruction. That is, they involve students in active learning and are interesting and engaging.

The authors of this book often work with elementary and secondary science teachers to design meaningful, integrated performance learning assessments. We always begin by asking teachers to consider what these

△ *Meaningful performance assessments allow teachers to assess concept understanding, process skills, and habits of mind through a cohesive set of activities.*

performances would look like—what a student might "do" as part of, or following, an integrated unit of instruction to show growth and development in understanding concepts, acquiring multidisciplinary skills, developing habits of mind, and connecting science to technology and society.

A meaningful performance task is an activity or series of activities designed to instruct and gather information concerning what the student knows or can do. Ideally, the performance task should not only teach, but should also assess one or more levels of concept understanding, the ability to perform or apply process skills, and habits of mind. The total performance task should include one or more relevant learning activities with opportunities to show learning through such things as carefully constructed notebook entries, criterion-referenced test items, and writing prompts.

We have identified 10 steps that we believe are critical to designing meaningful, integrated performance tasks. These steps are not necessarily sequential. Like juggling balls in the air, teachers need to keep all of the steps in motion while maintaining a clear perspective on standards and instructional goals.

1. Consider standards.

Review national, state, and local standards documents to identify the important goals that are relevant to the age and ability level of the students for whom the performance is being developed. Such a review might include *Science for All Americans, Benchmarks for Science Literacy, National Science Education Standards,* and *Curriculum and Evaluation Standards for School Mathematics,* as well as state standards documents.

2. Examine and discuss behaviors of scientists "in action."

How do scientists use concepts and skills in the work they do? What habits of mind or dispositions do they exhibit in their work?

Students can observe scientists at work through videotape or televised programs, or visits to museums, zoos, and other informal learning centers. They may invite a scientist or engineer from the community to visit their classroom for an informal discussion of the work he or she performs.

3. Relate science at your grade level to components of real-world science.

Reflect on how science at your grade level compares to components of science in the real world. Consider (1) science concepts, including the applications and connections of concepts to technology and society, and other areas of the curriculum; (2) science skills, both process and thinking skills; and (3) habits of mind displayed by working scientists. Identify important concepts, skills, and habits of mind that are part of your science program and would be meaningful to assess.

4. Design a context for what you want to assess.

Think about a context for assessing the important components of science that you have identified. The context may be an activity or a series of activities that relate to an investigation; a problem-based situation; a decision-making situation; the clarification of an issue; the development of a point of view; or the generation of a project, product, or invention.

The context should provide an opportunity for students to show any or all of the following related to science:

- Knowledge and comprehension of concepts, application of concepts, and connection of concepts to real-world contexts
- Ability to use process and complex thinking skills to solve problems
- Attitudes and behaviors that are valued by the scientific community

Consider important components of other disciplines to assess as part of the performance task. For example, components of language literacy might be assessed through an oral presentation related to an investigation or project. Such a performance might look as follows:

> Students will demonstrate knowledge and use of basic science vocabulary when explaining a dinosaur diorama they make that includes features of the habitat; types and availability of food, water, and space for the animal; and a model showing the unique physical characteristics of the dinosaur they researched. They will use a variety of references and resources leading to the development of their animal habitats and models. In an oral or written report, students will explain how they used evidence to make inferences concerning the extinction of their dinosaur. They will describe similarities and differences between extinct animals and animals that exist today.

5. Clarify and enrich the meaningful performance task.

As you design the performance task, build in as many opportunities as possible for students to show what they know and can do. Create a rough draft of the performance task, and then revisit the important concepts, skills, and habits of mind from science and other disciplines. Consider additional opportunities to enrich the task by including more of the important components. Consider and add a variety of ways to gather information for student learning, such as notebook entries including action plans, data tables, written explanations, graphics and visuals, criterion-referenced items, and writing prompts.

Δ *As you design the performance task, build in as many opportunities as possible for students to show what they know and can do.*

6. Identify the important elements of the performance task to assess.

Consider any prior learning or experience that is required so that students are prepared for the performance task. For the example given in Step 4, students might have studied the following:

- Reptiles as a group of animals; dinosaur types; characteristics of herbivores, carnivores, and omnivores
- Factors (climatic, ecological, etc.) affecting habitats, habitat–animal relationships, food chains, animal structure and function
- Strategies for researching information and conducting interviews

Identify the concepts, process and thinking skills, and habits of mind that can be taught, reinforced, or assessed through the performance task. Include connections to mathematics, technology, and society whenever possible. The set of identified elements constitutes the performance criteria—the dimensions of concepts, skills, and habits of mind that students will have an opportunity to learn, reinforce, practice, and demonstrate through the performance. The following is a list of dimensions that might be assessed in the example in Step 4:

Developing conceptual knowledge

The student does the following:

- Describes the physical properties of a dinosaur
- Explains the relationship between structure and function
- Shows and describes features of the model habitat
- Describes the probable food chain that existed for this animal
- Describes the evidence that informs scientists about this dinosaur

Using process and thinking skills

The student does the following:

- Uses a variety of resources for information
- Classifies the animal as an herbivore, a carnivore, or an omnivore and tells why
- Makes a scale model of the dinosaur and describes the ratios used
- Infers why this animal became extinct
- Demonstrates skills through an oral presentation

Exhibiting worthwhile habits of mind or dispositions

The student demonstrates the following:

- Curiosity and a desire for knowledge by engaging in research and learning activities eagerly and cooperatively
- Open-mindedness and a respect for evidence by explaining that what is known now about extinct species may change over time with additional research
- Cooperation by working with others to complete a task

Connecting science with technology and society (S-T-S)

The student does the following:

- Explains how knowledge of dinosaurs has changed over time and infers why
- Provides names of resources available in the local community or elsewhere for learning more about dinosaurs and the field of paleontology

You will probably not assess all of the possible dimensions within a single performance task. Thus, it is important to prioritize instructional goals and determine which dimensions are most appropriate to address and assess.

7. Determine what an acceptable performance looks like.

Teachers must carefully determine the precise indicators of learning that comprise an acceptable performance. Then these indicators need to be clearly described so that both the student and others understand the critical learning features of the performance task. Students may be able to assist in identifying some of these indicators of a quality performance. In fact, including students in the discussion about the key learning indicators provides an opportunity for them to ask questions that often lead to clarification.

8. Establish a rubric for learning and assessment.

Using the key elements of the task, list the learning dimensions and the indicators of learning dimensions that you are targeting through the performance. The learning dimensions are the concepts, skills, and habits of mind to be exhibited during the task. The indicators are what students will do in the task to show knowledge and understanding of concepts (e.g., identify and describe the components of the water cycle), and the application of skills (e.g., make a prediction; measure the mass of an object) and behaviors that exhibit habits of mind (e.g., communicate results of an investigation; show honesty and accuracy in recording data and information).

When creating a rubric for a performance task, list the indicators in the order they are done in each activity. For example, one activity might have indicators listed in this order:

- Describes a plan for investigating the question
- Develops a hypothesis to suggest how one variable will affect another
- Correctly measures the distance between points A and B
- Collects and records data

For a rubric that provides clear, meaningful information, we recommend using a simple recording scheme, such as this three-item scale: "complete," "almost," "not yet." By placing this simple notation next to each learning indicator, both students and teacher can easily understand what specific knowledge or skill needs attention.

By considering each learning indicator separately, you can document whether or not a student demonstrated concept understanding, skill acquisition, habits of mind, and an understanding of S-T-S connections throughout the performance. As explained in Chapter 3, when a rubric is developed listing each aspect of learning independently, that rubric is labeled analytic. We believe that using this type of rubric is most effective for learning and for providing meaningful feedback. Therefore, all of our examples employ this approach.

9. Communicate the performance task.

At this point, you will write a complete and accurate description of the performance task for your own understanding and for other teachers who might use the task for learning and assessment. The following framework can help you to write the description of your performance assessment.

Δ *By considering each indicator separately, you can document whether or not a student demonstrated concept understanding, skill acquisition, habits of mind, and S-T-S connections throughout the performance.*

For the overall performance, provide the following:
- Title of the performance task
- Rationale for the performance
  Provide a rationale for the learning dimensions you will address and tell why they are important. You might explain the reasons why you selected certain concepts, skills, and habits of mind. You might also include some beliefs you have about the teaching/learning process that determined your decisions. Considerations related to the curriculum and to the population of students might also be relevant. The rationale should include a brief description of the ways students will be learning through the task as well as the ways their learning will be assessed.
- Criterion-referenced test items and writing prompt
  Provide criterion tests or writing activities that allow students to immediately apply knowledge or demonstrate writing skills. We have provided criterion tests for each task as well as a writing prompt that authentically relates to the specific context of each performance. For example, in the Dog-Washing activity, students are asked to develop a brochure. In the Magnetic Force activity, students are asked to write a letter to a recycling center regarding the use of magnets in recycling.
- Analytic rubric

For each activity in the performance, provide the following:
- Title of activity
- Materials
  Provide a list of equipment and materials needed.
- Description of the activity
  Describe the tasks and purpose of the activity.
- Presentation of the activity
  Explain how to present the activity to students. Offer management suggestions and other information needed for conducting the activity.
- Student response sheets or notebooks.
  Provide data or response sheets that students will use to record their ideas, drawings, and such. Data or response sheets should include all appropriate information needed by students to conduct their investigation. The sheets may include the title of the activity, inquiry question(s), clear directions, data tables, graphs, charts, lines for writing, space for drawings, questions for processing and meaning-making, and space for summary statements and conclusions.

10. Field test the performance task and revise as needed.

Once you have developed a task that seems to have merit, pilot the performance task with students. Ask them to evaluate the task. Use the rubric you created to provide feedback to students about their learning or use the rubric for student self-assessment. Often, students will surprise you with their efforts, and you may find that the expectations you had when developing the rubric are different from those that students were able to meet. Use student comments and reactions to revise the activities in the performance task; use student work to clarify and revise the rubric.

AN INVITATION   ◼

We have attempted to present a rationale for designing meaningful, integrated performance learning and assessment tasks. In this book, we have provided examples or prototypes of what such performance tasks might look like in the classroom. Some of the examples are tied to specific topics in science, while others simply provide a framework that allows teachers to select the context for the performance.

The major impetus for the performance assessments presented in this text comes from *Benchmarks for Science Literacy* by the American Association for the Advancement of Science (1993) and the *National Science Education Standards* developed by the National Research Council (1996). Specific benchmark statements from *Benchmarks for Science Literacy* and specific standards from the *National Science Education Standards* are identified for each of the performance tasks. A summary of the specific statements and benchmarks addressed is found in Box 4.1. These statements correspond to the concept understanding, skills, habits of mind, and real-world relationships a student might be able to develop, enhance, and show through the performances.

In the performances, learning is assessed through a variety of strategies: student-centered activities, criterion-referenced tests, and writing prompts. The activities that comprise each performance are essential to the whole because they offer students multisensory experiences and opportunities to actively demonstrate their understandings and abilities in science.

To meet national, state, and local standards of excellence, the prototypical performance tasks we offer in this book include the following:

- Inviting, engaging, student-centered questions and activities about the physical world and our relationship to it
- Critical dimensions of science as described in *Science for All Americans* and *Benchmarks for Science Literacy* and *National Science Education Standards* that relate to unifying concepts and processes, thinking skills, and scientific habits of mind
- Rationales for the performances and suggestions for presenting activities
- Lesson plans for hands-on activities with materials lists and student response sheets
- Different types of questions requiring different types of thinking
- Sample criterion-referenced test items
- Writing prompts that connect science to technology and society
- Flexible analytic rubrics that can be tailored to individual needs

An analytic rubric is presented at the end of each performance assessment. Indicators of learning are listed in the order in which they occur in the activities. Although many indicators are listed, teachers are encouraged to select those indicators that are of most importance to their instructional goals. Other indicators may also be added.

Habits of mind that are inherent in the activities are not specifically listed. However, in each activity there are ample opportunities for students to demonstrate desirable behaviors such as perseverance, honesty, and the ability to work in a collaborative group.

The purpose of the writing prompt is to allow students to express their ideas in a more open and creative manner as they apply science to a

△ The activities that comprise each performance are essential to the whole because they offer students multisensory experiences and opportunities to actively demonstrate their understandings and abilities in science.

real-world context. Responses to the writing prompts can show concept understanding and application as well as, at times, the students' attitudes and values. We have provided suggested criteria for assessing each of the writing prompts. However, teachers are encouraged to identify indicators of language literacy that are important to them.

In addition to the writing prompt, a set of criterion-referenced, multiple-choice test items is provided that corresponds to the main concepts. These questions are clustered in groups of two or three, each focusing on a specific concept. The items are offered as samples that provide multiple opportunities for students to show they understand a concept. If a student gets two or all three of the items related to a particular concept correct, one might reasonably assume that the student knows this concept. An answer key for the items is provided in the Appendix.

We hope that our prototypical examples and explanations assist you in designing performance tasks for learning and assessment. To this end we, the authors, invite you to use these assessments, to edit them, and to make them an integral part of their diverse curricular contexts in your classroom. There is no greater pleasure for us than seeing these prototypical performances used and transformed through the creative talents of our fellow educators. We anticipate that this is but the beginning of an important revolution in the area of learning and formative assessment and that the future of science education will be filled with rich, meaningful, integrated performance tasks.

BOX 4.1
Standards and Benchmarks: Correlations in Performance Tasks

Chapter 5: Adventuring Back in Time

National Science Education Standards

- Fossils provide evidence about the plants and animals that lived long ago and the nature of the environment at that time. (p. 134)
- Fossils provide important evidence of how life and environmental conditions have changed. (p. 160)
- Extinction of a species occurs when the environment changes and the adaptive characteristics of a species are insufficient to allow its survival. Fossils indicate that many organisms that lived long ago are extinct. Extinction of a species is common; most of the species that have lived on the earth no longer exist. (p. 158)

Project 2061 Statements from Benchmarks for Science Literacy

- Fossils can be compared to one another and to living organisms according to their similarities and differences. Some organisms that lived long ago are similar to existing organisms, but some are quite different. (p. 123)
- Similarities among organisms are found in internal anatomical features, which can be used to infer the degree of relatedness among organisms. In classifying organisms, biologists consider details of internal and external structures to be more important than behavior or general appearance. (p. 104)
- Scale drawings show shapes and compare locations of things very different in size. (p. 223)
- Make sketches to aid in explaining procedures or ideas. (p. 296)
- Find and describe locations on maps with rectangular and polar coordinates. (p. 297)

Chapter 6: The Dog-Washing Business

National Science Education Standards

- Objects have many observable properties, including size, weight, shape, color, temperature, and the ability to react with other substances. Those properties can be measured using tools such as rulers, balances, and thermometers. (p. 127)
- Materials can exist in different states—solid, liquid, and gas. Some common materials, such as water, can be changed from one state to another by heating or cooling. (p. 127)
- Heat moves in predictable ways, flowing from warmer objects to cooler ones, until both reach the same temperature. (p. 135)
- Abilities of Technological Design (p. 165)

Project 2061 Statements from Benchmarks for Science Literacy

- When warmer things are put with cooler ones, the warm ones lose heat and the cool ones gain it until they are all at the same temperature. (p. 84)
- Graphs can show a variety of possible relationships between two variables. (p. 219)
- Practical reasoning, such as diagnosing or troubleshooting almost anything, may require many-step, branching logic. (p. 233)

Chapter 7: May the Force Be With You

National Science Education Standards

- Objects can be described by the properties of the materials from which they are made, and those properties can be used to separate or sort a group of objects or materials. (p. 127)
- Magnets attract and repel each other and certain kinds of other materials. (p. 127)

(Continued)

BOX 4.1 (Continued)

Project 2061 Statements from Benchmarks for Science Literacy

- Without touching them, a magnet pulls on all things made of iron and either pushes or pulls on other magnets. (p. 94)
- People can often learn about things around them by just observing those things carefully, but sometimes they can learn more by doing something to the things and noting what happens. (p. 10)
- Offer reasons for findings and consider reasons suggested by others. (p. 286)
- Make sketches to aid in explaining procedures or ideas. (p. 296)
- When people care about what's being counted or measured, it is important for them to say what the units are. (p. 292)

Chapter 8: Water, Water Everywhere

National Science Education Standards

- Water, which covers the majority of the earth's surface, circulates through the crust, oceans, and atmosphere in what is known as "the water cycle." Water evaporates from the earth's surface, rises and cools as it moves to higher elevations, condenses as rain or snow, and falls to the surface where it collects in lakes, oceans, soil, and in rocks underground. (p. 160)
- Water is a solvent. As it passes through the water cycle, it dissolves minerals and gases and carries them to the oceans. (p. 160)

Project 2061 Statements from Benchmarks for Science Literacy

- When liquid water disappears, it turns into a gas (vapor) in the air and can reappear as a liquid when cooled, or as a solid if cooled below the freezing point of water. Clouds and fog are made of tiny droplets of water. (p. 68)
- The cycling of water in and out of the atmosphere plays an important role in determining climatic patterns. Water evaporates from the surface of the earth, rises and cools, condenses into rain or snow, and falls again to the surface. The water falling on land collects in rivers and lakes, soil, and porous layers of rock, and much of it flows back into the ocean. (p. 69)
- Offer reasons for their findings and consider reasons suggested by others. (p. 286)
- Know why it is important in science to keep honest, clear, and accurate records. (p. 287)

Chapter 9: A Wholesome Partnership

National Science Education Standards

- Objects have many observable properties, including size, weight, shape, color, temperature, and the ability to react with other substances. (p. 127)
- Identify questions that can be answered through scientific investigations; design and conduct a scientific investigation; use appropriate tools and techniques to gather, analyze, and interpret data; develop descriptions, explanations, predictions, and models using evidence; and think critically and logically to make the relationships between evidence and explanations. (p. 145)

Project 2061 Statements from Benchmarks for Science Literacy

- No matter how parts of an object are assembled, the weight of the whole object is always the same as the sum of the parts; and when a thing is broken into parts, the parts have the same total weight as the original thing. (p. 77)
- Mathematical statements can be used to describe how one quantity changes when another changes. (p. 219)

Chapter 10: A-W-L for One and One for A-W-L (Air-Water-Land)

National Science Education Standards

- Abilities necessary to do scientific inquiry (pp. 122–123)
- Changes in environments can be natural or influenced by humans. Some changes are good, some are bad, and some are neither good nor bad. Pollution is a change in the environment that can influence the health, survival, or activities of organisms, including humans. (p. 140)
- Science cannot answer all questions and technology cannot solve all human problems or meet all human needs. (p. 169)

Project 2061 Statements from Benchmarks for Science Literacy

- Scientists do not pay much attention to claims about how something they know about works unless the claims are backed up with evidence that can be confirmed with a logical argument. (p. 11)
- Statistical predictions are typically better for how many of a group will experience something than for which members of the group will experience it—and better for how often something will happen than for exactly when. (p. 227)
- Students should keep records of their investigations and observations and not change the records later. (p. 286)
- Students should offer reasons for their findings and consider reasons suggested by others. (p. 286)

Chapter 11: The Mysterious Package

National Science Education Standards

- All animals depend on plants. Some animals eat plants for food. Other animals eat animals that eat the plants. (p 129)
- Populations of organisms can be categorized by the function they serve in an ecosystem. All animals, including humans, are consumers, which obtain food by eating other organisms. (pp. 137–138)

Project 2061 Statements from Benchmarks for Science Literacy

- A great variety of living things can be sorted into groups in many ways using various features to decide which things belong to which group. (p. 103)
- Almost all kinds of animals' food can be traced back to plants. (p. 119)
- Some source of "energy" is needed for all organisms to stay alive and grow. (p. 119)
- Two types of organisms may interact with one another in several ways: They may be in a producer/consumer, predator/prey, or parasite/host relationship. (p. 117)
- Tables and graphs can show how values of one quantity are related to values of another. (p. 218)
- The graphic display of numbers may help to show patterns such as trends, varying rates of change, gaps, or clusters. (p. 224)

Chapter 12: Up, Up, and Away

National Science Education Standards

- Identify questions; design and conduct investigations; use tools and techniques to gather, analyze, and interpret data; develop descriptions, explanations, predictions, and models using evidence; think critically and logically to make the connections between evidence and explanations; recognize and analyze alternative explanations and predictions; communicate scientific procedures and explanations; use mathematics in all aspects of scientific inquiry. (pp. 145 and 148)

(Continued)

BOX 4.1 (Continued)

Chapter 12: Up, Up, and Away

Project 2061 Statements from Benchmarks for Science Literacy

- Scientists differ greatly in what phenomena they study and how they go about their work. Although there is no fixed set of steps that all scientists follow, scientific investigations usually involve the collection of relevant evidence, the use of logical reasoning, and the application of imagination in devising hypotheses and explanations to make sense of the collected evidence. (p. 12)
- If more than one variable changes at the same time in an experiment, the outcome of the experiment may not be clearly attributable to any one of the variables. It may not always be possible to prevent outside variables from influencing the outcome of an investigation, but collaboration among investigators can often lead to research designs that are able to deal with such situations. (p. 12)
- Thinking about things as systems means looking for how every part relates to others. (p. 265)

Chapter 13: Picture Perfect Professions

National Science Education Standards

- Many people choose science as a career and devote their entire lives to studying it. Many people derive great pleasure from doing science. (p. 141)
- Women and men of various social and ethnic backgrounds—and with diverse interests, talents, qualities, and motivations—engage in the activities of science, engineering, and related fields such as the health professions. Some scientists work in teams, and some work alone, but all communicate extensively with others. (p. 170)
- Science requires different abilities, depending on such factors as the field of study and type of inquiry. Science is very much a human endeavor, and the work of science relies on the basic human qualities, such as reasoning, insight, energy, skill, and creativity—as well as on scientific habits of mind, such as intellectual honesty, tolerance of ambiguity, skepticism, and openness to new ideas. (p. 170)

Project 2061 Statements from Benchmarks for Science Literacy

- People can learn about others from direct experience, from the mass communications media, and from listening to other people talk about their work and their lives. (p. 154)
- Each culture has distinctive patterns of behavior, usually practiced by most of the people who grow up in it. (p. 155)
- As students begin to think about their own possible occupations, they should be introduced to the range of careers that involve technology and science, including engineering, architecture, and industrial design. Through projects, readings, field trips, and interviews, students can begin to develop a sense of the great variety of occupations related to technology and to science and what preparation they require. (p. 46)

Source: National Science Education Standards are reprinted with permission from **The National Science Education Standards** © **1997** by the National Academy of Sciences, courtesy of the National Academics Press, Washington, D.C. Benchmarks for Science Literacy are reprinted by permission of Oxford University Press, Inc.

PART II

Application

Adventuring Back in Time 5

Rationale

This set of performance tasks assumes that students are familiar with the basic concepts relating to fossils and paleontology. They should know that fossils are the remains of animals and plants that have been preserved naturally. Students should be familiar with casts and molds, imprints, and mineral-replaced remains.

For Activity 1, students will need to know how to make a scale drawing and be able to locate points on a grid by using the axis symbols that define the coordinates. If students lack experience making scale drawings, you may want to conduct this part of the performance as a teacher-directed activity. In Activity 2, students will be asked to measure in centimeters and to find the mass in grams. They will be asked to describe, compare, and infer, focusing on the characteristics that make it possible to identify a fossil and to distinguish it from others. Then, in Activity 3, students will compare similarities and differences among the features of their fossils and those of animals that are alive today.

National Science Education Standards

- Fossils provide evidence about the plants and animals that lived long ago and the nature of the environment at that time. (p. 134)
- Fossils provide important evidence of how life and environmental conditions have changed. (p. 160)
- Extinction of a species occurs when the environment changes and the adaptive characteristics of a species are insufficient to allow its survival. Fossils indicate that many organisms that lived long ago are extinct. Extinction of species is common; most of the species that have lived on the earth no longer exist. (p. 158)

Project 2061 Statements from Benchmarks for Science Literacy

- Fossils can be compared to one another and to living organisms according to their similarities and differences. Some organisms that lived long ago are similar to existing organisms, but some are quite different. (p. 123)
- Similarities among organisms are found in internal anatomical features, which can be used to infer the degree of relatedness among organisms. In classifying organisms, biologists consider details of internal and external structures to be more important than behavior or general appearance. (p. 104)
- Scale drawings show shapes and compare locations of things very different in size. (p. 223)
- Make sketches to aid in explaining procedures or ideas. (p. 296)
- Find and describe locations on maps with rectangular and polar coordinates. (p. 297)

Source: Reprinted with permission from **The National Science Education Standards** © 1997 by the National Academy of Sciences, Courtesy of the National Academies Press, Washington, D.C. Benchmarks for Science Literacy by permission of Oxford University Press, Inc.

CAN YOU DIG IT?

Description of Activity

Students will be given a fossil-rich "site." They will be asked to make a scale drawing that is one half the size of the site and to label the coordinates on a grid. Using simple tools, students will excavate four different fossils from the site. They will describe the fossils and use coordinate symbols to indicate the location of the specimens on the scale drawing. The four fossils may be buried in the pans in similar locations so that the location data will be the same for each student. Alternately, the four fossils can be buried in different locations in each pan, providing different data for each student group. Teachers may decide which situation is best for them.

Materials

- Large rectangular baking pans or cardboard trays filled with 2 to 3 centimeters of sand or cat litter (cardboard trays from canned cat food may be obtained from grocery stores—approximately 26 cm x 36 cm x 5 cm)
- Plastic spoons or other small digging tools
- 4 plastic (or real) fossils of crinoid stems, brachiopods, pelecypods, and petrified wood (or other fossils that are available)
- Fossil kits from Creative Dimensions, P.O. Box 1393, Bellingham, WA 98227 (optional)

Presenting the Activity

Tell students they have been invited by the local museum paleontologist to assist with an exploratory dig at a fossil-rich quarry. Their task is to assist the paleontologist in identifying the locations of fossils found at the site by making a scale drawing that is one half the size of the "site" (the box with the buried fossils). After measuring the site and making the map, students should mark off 2-centimeter sections along the top and the sides of the map to make a grid. They will label the top of the grid with letters and the sides with numbers. Figure 5.1 is an example of a site map developed by students. Students may create different variations of this map, but each map should be approximately one half the size of their actual site. Students will search for the fossils, describe them, and record the location where they were found on their maps.

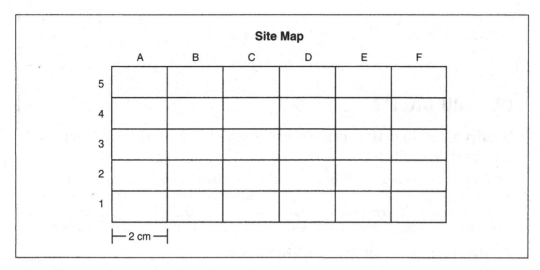

Figure 5.1

ACTIVITY 1

Name: _____

Date: _____

CAN YOU DIG IT?

Inquiry Question: How can we use a map to show the locations of our fossil finds?

For weeks you have prepared for your fossil dig with the paleontologist. Today, you arrive at the site to dig for fossils. The site is an abandoned sedimentary rock quarry that, over time, has been covered with loose material. Fossils are buried in the material. To begin, record observations, including a description of the site and the size of the area to be investigated.

Description (material, color, landscape features, etc.)

_____

Size of area (in centimeters)

_____

Make a scale drawing of the site that is one half the size of the actual area you are investigating, on a separate sheet of paper. Mark off the drawing in 2cm sections along the top and sides and draw the lines to create a grid of 2cm-by-2cm blocks. Label the grid with letters across the top or bottom (left to right) and numbers along the side.

You will use the grid to record the locations of your fossils. Identify fossil locations by letter and number. For example, C2 would be the third square from the left intersecting with the second one from the bottom or the top (depending on how you label your grid). Using the digging tools provided, carefully probe the quarry site until you find four fossils. Write a description of each fossil and give its coordinate location on the data table below.

	Description	Coordinates
W	_____	_____
X	_____	_____
Y	_____	_____
Z	_____	_____

Assign each fossil a letter: W, X, Y, or Z. Next put each letter on the grid (map) to show the exact locations at the site where each fossil was found.

Explain how your map can be used by another explorer to find the locations of your fossils.

_____

_____

_____

_____

_____

_____

ADDITIONAL LESSON IDEAS OF YOUR OWN

NOW, PICTURE THIS

Description of Activity

Students are asked to make a drawing of each of the four fossils. This is an important skill required of researchers to record the intricate features of their finds. Students will measure the length, width, and mass of each fossil. Then, they will study each fossil's characteristics to determine if any resemble organisms that are alive today.

Materials
• Fossils • Hand lenses • Drawing pencils • Metric rulers • Balances • Mass sets

Presenting the Activity

Inform students that scientists use words and pictures to record as much information as possible when doing research in the field. Tell them that careful observations, written descriptions, and drawings are used to record characteristics and unusual details of their finds. Finding and recording the size and the mass of a specimen are important, as well. As students identify their own fossil finds, stress that it is important to note whether the fossil resembles any plant or animal with which they are already familiar, such as organisms in the environment or those observed in books or in other media.

ACTIVITY 2

Name: _____

Date: _____

NOW, PICTURE THIS

Inquiry Question: Do some of the fossils look similar to organisms we see today?

The four specimens you excavated from your site are fossils of organisms that lived long ago. Draw each specimen in the space on the data table below and record details of its structure. Measure the length and width (at the longest and widest points) of each fossil and record the data. Use the balance to find the mass of each specimen in grams and record the data. If any of the fossils resemble currently existing organisms, make inferences as to what familiar organisms the fossils might be related to.

Drawing of Specimen	Length (cm)	Width (cm)	Mass (g)	Organisms Fossil Resembles

HERE'S LOOKING AT YOU, KID

Description of Activity

Comparing similarities and differences among fossils and between fossils and other known organisms gives scientists clues to evolutionary patterns.

Materials
• Fossils • Hand lenses • Reference books about fossils • Fossil kits (optional)

Presenting the Activity

Students are often interested in knowing the names of the fossils they find. Ask them to begin by comparing the fossils to one another. What characteristics do they have in common? Identify and discuss similarities and differences. Challenge students to find at least one way that two of the fossils are alike.

Provide field guides and reference books about fossils. Using these guides, students will identify the type or name of each fossil and record the information on their data sheets. From these resources and others, students may identify similar present-day organisms. Students should learn as much as possible about their four fossils and the organisms they resemble.

ACTIVITY 3   Name: _____

                                    Date: _____

HERE'S LOOKING AT YOU, KID

Inquiry Question: How do fossils compare to one another and to other organisms?

Comparing Specimens to One Another

Use a hand lens to study your specimens. Compare them and list one way any two are alike.

_____

_____

Comparing Specimens to Other Organisms

You may use reference books to help find the name of each fossil and the names of any present-day organisms that are similar to each fossil. Record the specimen names, the names of the organisms they resemble, and the characteristics of the fossils that are similar to the present-day organisms.

	Specimen Name	Similar Organism	Similar Characteristics
W	_____	_____	_____
X	_____	_____	_____
Y	_____	_____	_____
Z	_____	_____	_____

Extension Activity

What interesting information do scientists know about fossils? (for use with Creative Dimensions Fossil Kit or another fossil kit or additional fossils)

For each of the fossils you have, locate the information card or research information about the different fossils. Use the information to record the following data for each fossil.

Name	Age	Origin	Type of Fossil
_____	_____	_____	_____

Two Items of Interest

_____

_____

Name	Age	Origin	Type of Fossil
_____	_____	_____	_____

Two Items of Interest

_____

_____

Name	Age	Origin	Type of Fossil
_____	_____	_____	_____

Two Items of Interest

_____

_____

Name	Age	Origin	Type of Fossil
_____	_____	_____	_____

Two Items of Interest

_____

_____

Summarize: In what ways can information from the past help us understand the present?

_____

_____

_____

CRITERION-REFERENCED TEST: ADVENTURING BACK IN TIME

Answer the following multiple-choice questions related to the key concepts of the characteristics and composition of fossils and the relationships of fossils to present-day organisms.

1. We can learn about the earth and organisms that lived millions of years ago by studying
   a. living things
   b. sea shells
   c. fossils
   d. igneous rock

2. When scientists find a fossil, they compare it to
   a. present-day organisms only
   b. fossils only
   c. fossils and present-day organisms
   d. nothing else

3. Fossils allow us to study organisms that
   a. lived long ago
   b. lived recently
   c. are currently living
   d. never lived

4. Jane found a fossil that looked like a shell. She knew it was a fossil because it was made of
   a. human-made materials
   b. the same thing as shells
   c. sand
   d. minerals that replaced the shell

5. Fossils _____ look like organisms that exist today.
   a. never
   b. always
   c. sometimes

6. Scientists have learned a lot about how organisms evolved by studying
   a. fossils and present-day animals
   b. rock layers
   c. igneous rock
   d. present-day animals

7. Which of the elements below is NOT a type of fossil?
   a. casts
   b. sandstone
   c. molds
   d. petrified wood

8. When scientists find a fossilized piece of a dinosaur, they try to reconstruct the whole animal. They do this by
   a. comparing the fossil to the structure of other dinosaurs
   b. studying the rock it came from
   c. studying living organisms
   d. comparing the fossil to present-day dinosaurs

9. Plants and animals that exist today are
   a. similar to all fossils
   b. similar to some fossils and different from some fossils
   c. different from all fossils
   d. totally unlike fossils

WRITING PROMPT: MAKING A PHOTOGRAPHIC JOURNAL

Any expedition is a one-time experience; the same events will not likely be repeated. Once fossils are removed from a site, their position, structure, and location can never be accurately recorded again. It is critical that precise records are kept of the experience.

Part I: On your expedition to discover fossils, you were asked by the paleontologists to act as the photojournalist to capture the unique characteristics of the trip. The scientists asked for four entries in a journal, each consisting of a snapshot and a written explanation of the image. Use your imagination to create unique features of your expedition. (You are not limited to the actual excavation site or specimens from your previous experience. You may create a more interesting and exciting expedition for your photo journal.)

Entries in the journal might include the following:

- A picture and description of the site, complete with researchers and equipment
- Paleontologists in action showing the processes they go through to extract fossils from the earth
- The finding of a fossil that is similar to an animal that might be found on earth today
- Pictures and descriptions of unusual fossils found at the site
- Other unique and imaginative, but realistic features of your experience

For each journal entry, provide a title, a picture, and a detailed description of each picture using at least two sentences.

Part II: Write a detailed explanation or be prepared to give a demonstration on how the structures and features of fossils provide evidence that organisms that existed in the past are similar to organisms that exist on earth today.

Be prepared to show your photo journal at a National Convention of Paleontologists. Have fun!

Part I:

Description of photo: _____

_____

_____

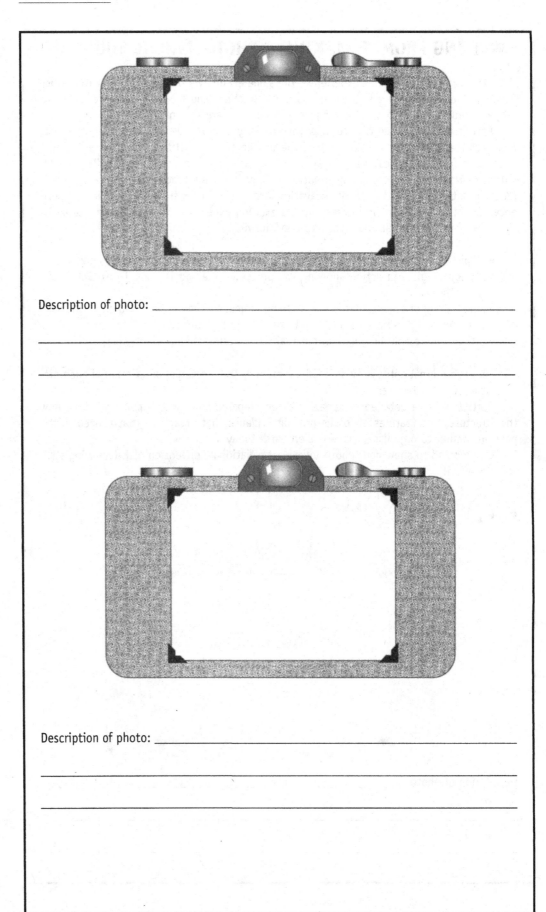

Description of photo: _____

_____

_____

Description of photo: _____

_____

_____

Description of photo: _____

_____

_____

Part II Explanation:

_____

_____

_____

ANALYTIC RUBRIC: ADVENTURING BACK IN TIME

Indicators of Learning Relate to These Unifying Concepts and Processes

- Systems, Order, and Organization
- Evidence, Models, and Explanations
- Constancy, Change, and Measurement
- Evolution and Equilibrium
- Form and Function
- Mathematics, Literacy, and Thinking Skills

(Items in parentheses below identify some of the concepts and skills assessed through the task.)

SCALE

	Complete	Almost	Not Yet	Comments
Activity 1: Can You Dig It? **The student:** • Wrote a detailed description of the site (Observation; Communication) • Measured the size of the site in centimeters correctly • Drew a model to scale (Made a Model) • Labeled the coordinates correctly • Described properties of four fossils accurately (Properties of Objects; Explanation) • Gave coordinates of locations of four fossils (Classification) • Placed fossils or letters correctly on the map (Graphing) • Gave logical answer regarding use of map (Explanation; Reasoning)				
Activity 2: Now, Picture This **The student:** • Drew or traced four specimens with detail (Observation; Creating Representations) • Measured and recorded length of each fossil • Measured and recorded width of each fossil • Measured and recorded mass of each fossil • Made inferences about fossil relationships to existing organisms (Evidence; Change; Evolution; Comparing; Reasoning)				

	Complete	Almost	Not Yet	Comments
Activity 3: Here's Looking at You, Kid **The student:** • Listed one way two fossils are alike (Comparing) • Found and recorded names of four fossils (Classification) • Identified organisms similar to fossils (Comparing; Making Analogies) • Listed characteristics of fossils that are similar to present day organisms (Properties of organisms)				
Criteria for Photographic Journal **The student:** **Part I** • Drew four pictures related to an expedition (Creating Representations; Inventing) • Gave a title and detailed description of each picture (two sentence minimum) • Linked classroom instruction to a science career area (Concept Application; Relationships between Science-Technology-Society; Meta-Cognition; Communication) **Part II** • Wrote a detailed explanation or gave a demonstration to explain how fossils provide evidence of relationships between organisms of the past and present (Concept Understanding; Explanation)				

Scoring Scale	
Complete	Student exhibited the indicator
Almost	Student showed some evidence that the indicator was exhibited, but something is incorrect or missing
Not Yet	Student did not show evidence of learning for that indicator

The Dog-Washing Business

6

Rationale

This performance assumes that students have experience reading thermometers in degrees Celsius and measuring the volume of liquids in milliliters.

Students will learn about the transfer of energy by observing change in temperatures in Activity 1. In Activity 2, students will explore changes in temperature over time, collect data, and make graphs to communicate their findings. They will also be asked to make predictions by extending the lines on their graphs (extrapolation). In Activity 3, students will use their new knowledge and problem-solving skills to develop an action plan for a business, applying what they learned in Activities 1 and 2 to a real-world situation. In doing so, they will be showing a link between science, technology, and society.

National Science Education Standards

- Objects have many observable properties, including size, weight, shape, color, temperature, and the ability to react with other substances. Those properties can be measured using tools, such as rulers, balances, and thermometers. (p. 127)
- Materials can exist in different states – solid, liquid, and gas. Some common materials, such as water, can be changed from one state to another by heating or cooling. (p. 127)
- Heat moves in predictable ways, flowing from warmer objects to cooler ones, until both reach the same temperature. (p.135)
- Abilities of Technological Design (p.165)

Project 2061 Statements from Benchmarks for Science Literacy

- When warmer things are put with cooler ones, the warm ones lose heat and the cool ones gain it until they are all at the same temperature. (p. 84)
- Graphs can show a variety of possible relationships between two variables. (p. 219)
- Practical reasoning, such as diagnosing or troubleshooting almost anything, may require many-step, branching logic. (p. 233)

Source: Reprinted with permission from **The National Science Education Standards** © 1997 by the National Academy of Sciences, Courtesy of the National Academies Press, Washington, D.C. Benchmarks for Science Literacy by permission of Oxford University Press, Inc.

THE HOT AND COLD OF DOG WASHING

Description of Activity

In this task, students will measure the volume of water in milliliters and measure the temperature in degrees Celsius. (Fahrenheit equivalents are not necessary as all the relative information is given.) They will discover what happens to water temperature when cold water is mixed with hot water. Hot water should be at least 40°C (between 100 and 110°F); it may be necessary to heat the water using a hot plate.

Safety

Be sure that the water and beakers are not too hot for students to handle. Pot holders and safety goggles should be used when hot water is in use. Ice water or cold water from the tap may be used for the cold water.

In processing the results, students will analyze data to determine if the water gained heat or lost heat and refer to changes as a transfer of energy.

Materials
• Student thermometers measuring degrees Celsius • Beakers with mL markings or plain beakers or cups and a graduated cylinder with mL markings • Hot plate for heating water (optional) • Pot holders and safety goggles

Presenting the Activity

Explain to students that when a person offers a service or starts a business, there are many issues to consider. For example, if Willie and Anita plan to open a dog-washing and -grooming business, what are some of the things they will need to consider? Aside from equipment purchases, location considerations, and financing, point out the need for the availability of water at a suitable temperature for washing the dogs. Have students assume that the available water is too cold for a comfortable bath. Students will assist Willie and Anita with their need to find out how to get the water to a comfortable temperature for washing the dogs. Thus, students will investigate the transfer of heat energy by finding out what happens when hot water is added to cold water. They will use this information to help Willie and Anita plan their business strategy.

ACTIVITY 1

Name: _____

Date: _____

THE HOT AND COLD OF DOG WASHING

Inquiry Question: What happens when hot water is mixed with cold water?

Willie and Anita want to start a dog-washing and -grooming business to earn money to go to camp. Since the water they have available is rather cold, they need to investigate ways to warm the water before washing the dogs. They have asked you to help them learn about the transfer of heat energy so they can get the proper water temperature needed to provide dog-washing services. They have asked your help in doing the following investigation.

△ *Safety Check:*
Do you have your pot holders and safety goggles ready? You'll need these for handling hot water and beakers.

Measure 200 mL of cold water into one beaker (A) and 200 mL of hot water into a second beaker (B). Using a thermometer, measure the temperature of the water in each beaker in degrees Celsius. Record data:

Beaker A _____ °C   Beaker B _____ °C

If the contents of Beaker A were poured into Beaker B, predict what the temperature of the water will be in degrees Celsius.

I think the temperature will be _____

because_____

Now, pour the contents of Beaker A into Beaker B. Stir. Record the temperature of the water.

The temperature of the water is_____

How did your prediction compare to the actual temperature?

_____

_____

Consider the water that was in Beaker A. Did it lose or gain heat after it was mixed with the water in Beaker B? _____

Consider the water that was in Beaker B. Did it lose or gain heat after it was mixed with the water in Beaker A? _____

What conclusion can you draw?

_____

_____

Based on what you learned, what advice can you give Willie and Anita that might help them plan their business?

_____

_____

ADDITIONAL LESSON IDEAS OF YOUR OWN

TIME AND TEMPERATURE

Description of Activity

Students will use the data they have collected about the temperature of the water in Beakers A and B to answer the inquiry question: What will happen to the temperature of the water in 30 minutes?

This activity will need to be done immediately following Activity 1 since heat loss will occur right away, or Activity 1 can be repeated as part of Activity 2. Students should have experience noting time and collecting data using a thermometer. In this investigation, they will record data and make a line graph to show changes in temperature over time.

You may wish to provide a graph, or students may construct their own graphs. Make sure both axes on the graph are labeled properly and that the points are connected by a smooth line. Students will make predictions about what the temperature will be after 40 minutes based on the curve of the line that connects the points (extrapolation).

Materials

- Student thermometers measuring degrees Celsius
- Beakers of water (one marked A, one marked B)
- Clock with minute hand
- Safety goggles

Presenting the Activity

Remind students that Willie and Anita need to keep the water at an appropriate temperature in order to give dog baths. They wonder what will happen to the temperature of the water over time. Willie and Anita ask your students to investigate what will happen to the temperature of the water in 30 minutes. Will it gain heat, will it lose heat, or will it stay the same? To answer the questions, students can immediately take the temperature of the mixture of water in the beaker and record it and then take temperature readings at 10 minutes, 20 minutes, and 30 minutes. Or, if too much time has elapsed, they can repeat Activity 1 to get an initial reading and continue to take temperature readings every 10 minutes.

Students should make a prediction, conduct the test, collect data, and then graph their data. A graph can be provided or created. Temperature is plotted on the vertical axis and time is plotted on the horizontal axis (see Figure 6.1). Since the data show a loss of heat over time, the line connecting the points will be "downhill." Ask students to predict what the temperature will be after 40 minutes based on their investigation. Students should show their prediction on the graph by extending the line with a broken line and placing a point on the line at the 40-minute mark.

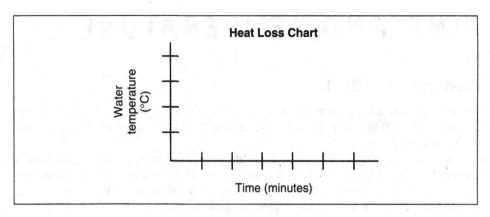

Figure 6.1

Aᴄᴛɪᴠɪᴛʏ **2**

Name: _____

Date: _____

TIME AND TEMPERATURE

Inquiry Question: What will happen to the temperature of the water after 30 minutes?

In the previous activity, you found the temperature of the water when the contents of Beaker A and Beaker B were mixed. Do you think the temperature of the water will change after 10 minutes? After 20 minutes? After 30 minutes? Will the water gain heat, lose heat, or stay the same?

Predict what the temperature of the water will be after 30 minutes.

I predict it will be _____

because _____ .

Record the temperature of the water. Then record the temperatures after 10 minutes, 20 minutes, and 30 minutes. Record the water temperatures on the data table.

Time	Temperature	Loss or Gain of Heat
Present		
10 minutes	_____	#1_____
20 minutes	_____	#2_____
30 minutes	_____	#3_____

Predict what the temperature of the water will be after 40 minutes.

I predict it will be _____

because _____ .

Now, make a temperature and time graph in the space provided below. Connect the three temperature points with a smooth line. Insert a dot at the 40-minute mark that corresponds to your prediction of the temperature. Show an extension of the line to this point using a broken line.

ADDITIONAL LESSON IDEAS OF YOUR OWN

SETTING UP BUSINESS

Description of Activity

Students will be given some information and asked to help develop a reasonable plan for arriving at a useful water temperature. Since this is a problem-solving activity, results may vary. Some students may use what they discovered in Activity 1, and some may not. Students may want to experiment with water using the beakers and thermometers, as they did in Activity 1. Or they may prefer to make paper-and-pencil calculations. If the temperature of the cold water is 20°C, it will allow for an easier mathematical solution to the problem.

Materials

- Student thermometers measuring degrees Celsius
- Beakers with mL markings, or plain beakers or cups and a graduated cylinder with mL markings (same as Activity 1)
- Safety goggles

Presenting the Activity

Anita and Willie need help developing a plan to get the water to a comfortable temperature for washing the dogs. The tap water is too cold. Their dad will provide hot water, but used alone, it is too hot for washing the dogs. Anita and Willie must come up with a plan for what they will do to get water to a useful temperature. The plan will require thought and calculation.

You may want to allow students to brainstorm criteria for a "good" plan. They should make note of each criterion. If students are not able to decide on the criteria for a good plan, you might provide a set of criteria to include logical ideas, use of a thermometer, collected data, a graph of data, and a reasonable solution. Throughout this discussion of a good plan, students should refer to water gaining and losing heat as the "transfer of energy."

ACTIVITY 3   Name: _____

                                  Date: _____

SETTING UP BUSINESS

Willie and Anita decide they need water that is about 30°C to comfortably wash the dogs. Their tap water is only 10°C. This water needs to gain heat to be useful. Their dad told them he will provide hot water from their home. The temperature of this water is 40°C. This water needs to lose heat to be useful.

Use the information you learned in Activity 1 to develop a plan for Willie and Anita so that they can take advantage of their dad's offer. Write an action plan in the space below.

_____

_____

_____

_____

_____

_____

_____

_____

_____

CRITERION-REFERENCED TEST: THE DOG-WASHING BUSINESS

Answer the following multiple-choice questions related to the key concept of the transfer of energy.

1. When ice cream is taken from the freezer and left on the counter, it
   a. gains heat from the room
   b. loses cold to the room
   c. gets colder the longer it is out
   d. does not change temperature

2. José takes a mug of hot chocolate outside on a cold day. In a short time, he discovers that the liquid in the mug
   a. loses cold
   b. loses heat
   c. gains heat
   d. gains cold

3. Sometimes when it is cold, Mary takes a hot-water bottle to bed with her. In about 2 hours, she observes that the bottle has changed temperature. Most likely it
   a. lost heat
   b. gained heat
   c. lost cold
   d. gained cold

4. During the winter, heat is added to a room to warm it. If the heat were turned off, what would happen to the air in the room? It would
   a. lose cold
   b. gain cold
   c. lose heat
   d. gain heat

5. The process of "thawing out" means the object must.
   a. gain heat
   b. gain cold
   c. lose heat
   d. lose cold

6. During the winter, Dan takes care of the animals in the barn. He adds a half bucket of warm water to the half bucket of cold water in the barn. The full bucket of water would then be
   a. the same temperature as it was
   b. colder than it was
   c. warmer than it was

WRITING PROMPT: ADVERTISING OUR BUSINESS

Willie and Anita have permission to distribute flyers advertising their business to residents of their neighborhood. As an advertising and graphic-design specialist, you have been called to assist them in developing a flyer that would convince customers that their service is needed and is better than other dog-washing and -grooming businesses in the area. Here are some questions and items you might consider for the flyer:

- What services do Willie and Anita offer?
- Why would dogs and their owners like these services?
- How will they ensure that dogs get a comfortable bath?
- Quotes from dogs; quotes from dog owners
- Information about Willie and Anita
- Pictures of qualified staff in action
- Pictures of satisfied customers
- Competitive costs

The flyer should be created using one 8.5 x 11-inch sheet of paper. The flyer may be organized, folded, colored, and created any way you wish. Display your flyer and be prepared to explain it to others.

ANALYTIC RUBRIC: THE DOG-WASHING BUSINESS

Indicators of Learning Relate to These Unifying Concepts and Processes

- Systems, Order, and Organization
- Evidence, Models, and Explanations
- Constancy, Change, and Measurement
- Evolution and Equilibrium
- Form and Function
- Mathematics, Literacy, and Thinking Skills

(Items in parentheses identify some of the concepts and skills assessed through the task.)

SCALE

	Complete	Almost	Not Yet	Comments
Activity 1: The Hot and Cold of Dog Washing **The student:** • Measured 200 mL of water accurately • Measured the temperatures of the hot and cold water accurately • Predicted the temperature of the mixture • Measured the temperature of the mixture accurately • Compared the prediction to the actual temperature (Comparing) • Correctly described what happened to the water in Beaker A (Observation; Explanation) • Correctly described what happened to the water in Beaker B (Observation; Explanation) • Drew a logical conclusion based on data (Logical Reasoning) • Gave sound advice showing understanding of energy transfer (Concept Understanding)				

	Complete	Almost	Not Yet	Comments
Activity 2: Time and Temperature **The student:** • Predicted with logical explanation • Collected and recorded data • Accurately recorded loss of heat • Made a second prediction with logical explanation • Constructed an appropriate line graph (Order and Organization) • Labeled both axes correctly (Graphing) • Showed a prediction on the graph (Concept Understanding)				
Activity 3: Setting Up Business **The student:** • Solved the problem by developing a logical plan (Reasoning and Problem Solving) (Additional student-generated indicators may be added here)				
Criteria for Advertising Our Business (Students may help determine criteria) The flyer should: • Describe the business accurately • Be attractive • Be informative • Be convincing (Descriptive Writing; Concept Application; Explanation and Representation)				

Scoring Scale	
Complete	Student exhibited the indicator
Almost	Student showed some evidence that the indicator was exhibited, but something is incorrect or missing
Not Yet	Student did not show evidence of learning for that indicator

May the Force Be With You 7

Rationale

This performance assumes that students are familiar with magnets and have done some preliminary work investigating magnetic properties of different objects. Students should know that metal objects are attracted to magnets, as opposed to plastic, glass, or wood.

In Activity 1, students will learn that not all metal objects are magnetic and will make inferences about metal objects that exhibit magnetic properties. In Activity 2, students will be asked to design a test to determine how far objects can be from a magnet and still be attracted to it. This demonstrates their ability to measure and express an understanding of a magnetic field. In Activity 3, students will set up a test to determine what happens when the poles of magnets come together, to show their understanding of the effect magnets have on one another. The activity ends with an opportunity for students to apply their understanding of forces within a field.

National Science Education Standards

- Objects can be described by the properties of the materials from which they are made, and those properties can be used to separate or sort a group of objects or materials. (p. 127)
- Magnets attract and repel each other and certain kinds of other materials. (p. 127)

Project 2061 Statements from Benchmarks for Science Literacy

- Without touching them, a magnet pulls on all things made of iron and either pushes or pulls on other magnets. (p. 94)
- People can often learn about things around them by just observing those things carefully, but sometimes they can learn more by doing something to the things and noting what happens. (p. 10)
- Offer reasons for findings and consider reasons suggested by others. (p. 286)
- Make sketches to aid in explaining procedures or ideas. (p. 296)
- When people care about what's being counted or measured, it is important for them to say what the units are. (p. 292)

Source: Reprinted with permission from **The National Science Education Standards** © **1997** by the National Academy of Sciences, Courtesy of the National Academies Press, Washington, D.C. Benchmarks for Science Literacy by permission of Oxford University Press, Inc.

METALS THAT ATTRACT

Description of Activity

Students are given a variety of metallic objects and a magnet. They should test each object to determine if it is attracted to the magnet. Students will record data by using an X to show which objects are attracted and will explain their observations.

Materials
• A piece of aluminum foil • A penny • A nickel • A paper clip • A brass fastener • A safety or straight pin • A bottle cap • Bar, ring, or horseshoe magnets

Presenting the Activity

In this activity, students will examine a variety of metallic objects to determine if all are attracted to a magnet. Instruct students to read the directions on the student data pages, collect data, and describe their observations. It is important that the inferences students make are based on their observations. Thus, if students observe that not all metallic objects are magnetic, they might infer that only certain metals are attracted to magnets. Students should realize that findings, such as "not all metals are magnetic," are what lead scientists to ask new questions about the properties of materials. Their understanding of the usefulness of magnets in the personal and social world is the focus of the open-ended question that addresses the relationships between science, technology, and society (S-T-S).

ACTIVITY **1**   Name: _____

                                        Date: _____

METALS THAT ATTRACT

Inquiry Question: Are all metal objects attracted to magnets?

I think _____

because _____.

   You will be given a set of objects that are made of different types of metal. Test the objects to see if they are magnetic. Place an X next to the object(s) that are attracted by the magnet.

Object	Attracted? (X = yes)
Aluminum foil	_____
Penny	_____
Nickel	_____
Paper clip	_____
Brass fastener	_____
Pin	_____
Bottle cap	_____

Describe your observations:

_____
_____
_____

Based on your observations, what inference(s) can you make about the objects that are attracted to magnets?

_____
_____
_____
_____
_____

Think of an example of how a magnet might be helpful in accomplishing a task. Describe the task and explain how the magnet would help:

_____
_____
_____
_____
_____
_____
_____
_____
_____
_____
_____

ADDITIONAL LESSON IDEAS OF YOUR OWN

CLOSE ENCOUNTERS

Description of Activity

In this task, students are asked to design an experiment to test whether objects must be in direct contact with a magnet to be affected by it. Students will make a prediction, describe a test or show one in a diagram, and perform the experiment. They are asked to design their own data tables and report the results. If students are not yet ready to design their own data tables, teachers may assist in developing an appropriate data table or the class may decide on a single test and data table that all students will use.

Materials

- Bar, ring, or horseshoe magnets
- Metal objects that are attracted to magnets from Activity 1
- Metric rulers or tape measures

Presenting the Activity

Explain to students that they should follow the directions on the student data sheet and make a prediction with an explanation. Tell them that they will then conduct the test to answer the inquiry question, record their data, draw a conclusion, and describe a magnetic field. Students will either design their own tests and create their own data tables or conduct the test and use the data table suggested by the class.

ACTIVITY 2   Name: _____

                                           Date: _____

CLOSE ENCOUNTERS

Inquiry Question: How far can an object be from a magnet and still be attracted to it?

I think _____

because _____.

   Using the magnet and the metal objects, design a specific test that would answer the above question. Describe the test by writing or drawing it in the space provided.

Perform the test. Develop a data table to show your findings. Draw the table below. (If the distance you measure is less than 1 centimeter, use the symbol <. Example: For a measurement of 1/2 a centimeter, write < 1 cm to represent less than 1 cm on the table.

Write a conclusion:

_____

_____

_____

_____

_____

_____

_____

   Based on what you have observed, draw what the magnetic field might look like for the test you conducted.

COMING ATTRACTIONS

Description of Activity

In this activity, students will investigate what happens when the poles of two magnets come together in the following combinations: N–N, N–S, and S–S. Students should conclude that like poles repel while opposite poles attract. Students should also determine that objects do not have to be touching the source of the force acting on them in order for the force to exert a push or pull. The open-ended question at the end of the activity asks students to think about forces that affect people in their everyday lives. The concept of force applies not only to magnetic force but also to other forces, such as gravitational, buoyant, electrical, and centripetal forces. In their examples, students may also describe forces that are applied by humans.

Materials
• 2 bar or ring magnets per person or team (If you are using ring magnets, put a red sticker on the N poles and a blue sticker on the S poles of the magnets.) • Metric rulers or tape measures • String • Scissors

Presenting the Activity

Ask students if they think magnets have an effect on one another. Have them discuss whether they think the poles are different and what happens when the poles of magnets come together. Instruct them to read the directions for the investigation and complete the data table and questions.

For the open-ended question at the end of the activity, tell students to think of other forces that affect people or objects. You may relate ideas to other forces students have studied in science or social studies, but allow students to give the examples. Students may describe natural forces such as gravity, or they may think about forces in a social science context as something that affects behavior, such as a force exerted by law, police, or gangs.

ACTIVITY 3

Name: _____

Date: _____

COMING ATTRACTIONS

Inquiry Question: What happens when the poles of two magnets come together in the following combinations: N–N, N–S, S–S?

Identify the north and south poles of a bar magnet (or red/blue sides of a ring magnet).

Place the north pole (red side) of one magnet near the north pole (red side) of the other magnet. Record your observations below.

_____

_____

_____

_____

Place the north pole (red side) of the first magnet near the south pole (blue side) of the second magnet. Record your observations below.

_____

_____

_____

_____

Finally, place the south pole (blue side) of the first magnet near the south pole (blue side) of the second magnet. Record your observations below.

_____

_____

_____

_____

Summarize your findings.

Magnetic poles	Observations (attract = X; repel = 0)
North–North (red–red)	
North–South (red–blue)	
South–South (blue–blue)	

What conclusion can you draw about the behavior of the magnets?

_____

_____

_____

How is a magnet a type of force?

_____

_____

_____

List two examples of forces in the environment. Describe each force and tell how each affects people or objects. Demonstrate or draw pictures of the forces, if possible.

_____

_____

_____

_____

_____

_____

_____

_____

_____

_____

_____

_____

_____

CRITERION-REFERENCED TEST: MAY THE FORCE BE WITH YOU

Answer the following multiple-choice questions related to the key concept of magnetic properties.

1. Which of the following objects will be attracted by a magnet?
   a. plastic cubes
   b. paper clips
   c. rubber bands
   d. paper strips

2. An object that pushes or pulls on another produces a
   a. force
   b. chemical
   c. change
   d. concept

3. A magnet produces a type of force because it
   a. has no effect on objects
   b. pulls on glass objects
   c. pushes or pulls on objects
   d. pushes on glass objects

4. When two magnetic poles are put close together, the result is
   a. gravitational pull
   b. no affect
   c. spinning of the magnets
   d. a push or pull

5. In order for a magnet to attract an object, it must be made of
   a. rubber
   b. plastic
   c. iron
   d. glass

6. Which diagram shows the correct way to put two bar magnets away so that the two ends will attract?

a.
N	S
S	N

b.
N	S
N	S

c.
S	N	N	S

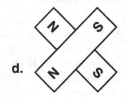
d.

7. A magnet has a north pole and a south pole. Two bar magnets are hung by strings so that the two north poles are facing one another.

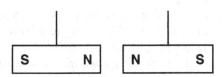

   In this situation, the magnets will
   a. push away from one another
   b. pull toward one another
   c. spin
   d. have no effect on one another

8. When Mark put the south pole of one magnet near the north pole of another magnet, he found that they
   a. pushed away
   b. repelled
   c. attracted
   d. had no effect

9. Sally put a strong magnet on the table near a compass that shows direction. She observed that the needle on the compass moved. She inferred that the needle of the compass was made of
   a. iron
   b. plastic
   c. glass
   d. aluminum

WRITING PROMPT: A FRIEND IN NEED

Your cousin Terry is in charge of an aluminum recycling project. She has been given donations of all sorts of metal objects including some aluminum. She knows she has to separate the aluminum from the other metals but does not have an effective way to do this.

Since you had such an important role in helping design the first space robot that went to the moon, you have been asked to help your cousin design a robot to help sort the metal for the recycling project. Use what you have learned about magnets and metals in your design. Draw a picture of what the Metal Monster would look like. Of course, you will need an instruction manual to go along with the robot so that your cousin will know how to operate it. Keep the manual short—no more than two pages. It should include step-by-step instructions and may include pictures.

ANALYTIC RUBRIC: MAY THE FORCE BE WITH YOU

Indicators of Learning Relate to These Unifying Concepts and Processes

- Systems, Order, and Organization
- Evidence, Models, and Explanations
- Constancy, Change, and Measurement
- Evolution and Equilibrium
- Form and Function
- Mathematics, Literacy, and Thinking Skills

(Items in parentheses identify some of the concepts and skills assessed through the task.)

SCALE

	Complete	Almost	Not Yet	Comments
Activity 1: Metals That Attract **The student:** • Made a logical prediction and gave a reason • Recorded data for magnetic items (paper clip, brass fastener, pin) (Observation; Classification) • Observed that all metal objects are not magnetic (Observation; Explanation) • Inferred that all metal objects are not made of the same materials (student might know that some are made of iron and some of aluminum) • Gave an example of the usefulness of magnets (Concept Understanding and Application; S-T-S) • Explained how a magnet can be used to solve a problem (Problem Solving and Reasoning)				
Activity 2: Close Encounters **The student:** • Made a logical prediction and gave a reason • Described a test in words or drawings (Explanation; Representation; Communication) • Performed the test and showed a data table with reasonable measurements (Measurement; Reasoning) • Used symbols properly, if applicable • Wrote a logical conclusion to show concept understanding (Systems, Order, and Organization)				

	Complete	Almost	Not Yet	Comments
• Drew the magnet showing lines of force with limited range (possible magnetic field) (Form and Function; Models and Explanation)				
Activity 3: Coming Attractions **The student:** • Recorded observations for N-N • Recorded observations for N-S • Recorded observations for S-S (Observation; Record Data) • Summarized findings (Models and Explanation) • Drew a conclusion about the behavior of magnets (Evidence; Reasoning; Communication) • Described magnetic force (Concept Understanding; Explanation) • Identified two examples of forces in the environment (Concept Understanding and Application; S-T-S) • Described how forces affect people or objects (Concept Application)				
Criteria for A Friend in Need The instruction manual should: • Be organized with a sequence of logical steps • Include instructions that show an understanding of magnetic attraction (Form and Function) • (Optional) Include pictures that fit the sequence of steps and aid in understanding how to use the "robot" to help solve the problem (Concept Understanding and Application; S-T-S; Representation; Models and Explanation)				

Scoring Scale	
Complete	Student exhibited the indicator
Almost	Student showed some evidence that the indicator was exhibited, but something is incorrect or missing
Not Yet	Student did not show evidence of learning for that indicator

Copyright © 2008 by Corwin Press. All rights reserved. Reprinted from *Integrating Science With Mathematics & Literacy: New Visions for Learning and Assessment*, 2nd. ed., by Elizabeth Hammerman and Diann Musial. Thousand Oaks, CA: Corwin Press, www.corwinpress.com. Reproduction authorized only for the local school site or nonprofit organization that has purchased this book.

Water, Water Everywhere 8

Rationale

Students can see water as rain, snow, and ice, but evaporation—the changing of water into invisible water vapor—is less noticeable. Activities in which students observe the disappearance of water from a dish or from a puddle help to define and reinforce their understanding of this natural phenomenon. Condensation, the formation of droplets of water on particles in the atmosphere or on surface objects (i.e., dew on grass, frost on a window), allows students to again observe water in a visible form. Clouds and fog are visible, but do students realize these are made up of tiny droplets of water? Students should be aware of the water cycle in nature and be able to trace it through its various stages.

Activity 1 allows students to observe the water cycle over time in a closed environment. Because the water cannot escape, it is present within the closed environment as both a liquid (visible stage) and a gas (invisible stage). Students must make inferences from observations for the stage they cannot observe and draw logical conclusions about the process as a whole.

In Activity 2, students will use the vocabulary of evaporation, condensation, and accumulation to describe their observations and inferences and will describe what precipitation looks like. They will explain the water cycle in words and pictures.

In Activity 3, students will discover that only water in its pure form goes through the water cycle. Salt and other minerals in water

National Science Education Standards

- Water, which covers the majority of the earth's surface, circulates through the crust, oceans, and atmosphere in what is known as "the water cycle." Water evaporates from the earth's surface, rises and cools as it moves to higher elevations, condenses as rain or snow, and falls to the surface where it collects in lakes, oceans, soil, and in rocks underground. (p. 160)
- Water is a solvent. As it passes through the water cycle, it dissolves minerals and gases and carries them to the oceans. (p. 160)

Project 2061 Statements from Benchmarks for Science Literacy

- When liquid water disappears, it turns into a gas (vapor) in the air and can reappear as a liquid when cooled, or as a solid if cooled below the freezing point of water. Clouds and fog are made of tiny droplets of water. (p. 68)
- The cycling of water in and out of the atmosphere plays an important role in determining climatic patterns. Water evaporates from the surface of the earth, rises and cools, condenses into rain or snow, and falls again to the surface. The water falling on land collects in rivers and lakes, soil and porous layers of rock, and much of it flows back into the ocean. (p. 69)
- Offer reasons for their findings and consider reasons suggested by others. (p. 286)
- Know why it is important in science to keep honest, clear, and accurate records. (p. 287)

Source: Reprinted with permission from **The National Science Education Standards** © 1997 by the National Academy of Sciences, Courtesy of the National Academies Press, Washington, D.C. Benchmarks for Science Literacy by permission of Oxford University Press, Inc.

are left behind when water evaporates, condenses, and accumulates as liquid on the bottom of the Ziploc bag. Students will apply the findings to the natural cycle and conclude that the water cycle is a distilling process. They will also research acid rain as a science-technology-society (S-T-S) connection. This performance task may be linked to units of study dealing with habitats and ecosystems, oceans, fresh water, and environmental education.

WATER CYCLE OBSERVATIONS

Description of Activity

This activity gives students an opportunity to make observations of a water cycle in action over 4 days. (You may also want to set up Activity 3 at this time.) The students will set up closed systems using Ziploc bags and be responsible for making their own observations throughout the week. Students will need access to a window where they will place their Ziploc bags. If your classroom does not have windows, locate an area in another classroom that may be used for this activity. The total amount of water added to the "system" should remain the same after all the water has gone through a cycle of evaporation, condensation, and accumulation at the bottom of the Ziploc bag. Data will be recorded and inferences will be made.

Materials
• Graduated cylinders • Small plastic cups (3 oz.) • Large Ziploc bags • Tape

Presenting the Activity

There are a variety of contexts to which this task may be applied. For example, it could be part of a study of the rainforest. Students would learn that these mysterious places occupy only about 8% of the earth's land surface but contain 40% to 50% of the world's plants and animals. Five million species of plants, insects, and animals live in the rainforests, but only one fourth of these have been identified and studied. About 80% of the world's insects live in the tropical rainforests, but many species are disappearing and may never be known to humans. Each year, an area of tropical rainforest the size of the state of Indiana is destroyed. Rainforests capture, store, and recycle water, which prevents floods, droughts, and erosion of the soil. In an effort to understand more about the rainforests' contribution to water purification, students may decide to study more about the way water is cycled through the natural system.

ACTIVITY 1

Name: _____

Date: _____

WATER CYCLE OBSERVATIONS

Inquiry Question: What happens to water when it is left in a warm, closed system such as inside a Ziploc bag?

Tropical rainforests are home to 5 million species of plants, insects, and animals, many of which will become extinct before humans can ever discover them. In this type of ecosystem, water is a plentiful and valuable resource. The rainforests capture, store, and recycle water, which prevents floods, droughts, and erosion of the soil. In order to understand more about the rainforests' contribution to water purification, you can study what happens to water in a closed environment or system.

Using the graduated cylinder, measure 40 mL of water into a small cup. Place the cup in the corner of a Ziploc bag. Tape the cup to the bag to hold it in place. Zip the bag shut and tape it to a window where it will get sunlight and heat.

Predict what will happen to the water inside the cup over time.

I predict _____

because _____.

Observe the bag each day for 4 days and record observations on the chart below.

Day	Date/Time	Observations
1		
2		
3		
4		

At the end of Day 4, write a summary of what happened to the water in the cup.

_____

_____

Estimate or measure how much water is in the Ziploc bag. Record your measurement and compare it to the amount of water you put into the bag.

_____

_____

ADDITIONAL LESSON IDEAS OF YOUR OWN

COMPARING WATER CYCLES

Description of Activity

Students will use the vocabulary related to the water cycle—evaporation, condensation, accumulation, and precipitation—to describe the various phases they observed in the closed system. They will describe the evidence they observed and the inferences they made for each of the stages. Students will use words and pictures to describe the water cycle in the closed system.

Materials
• None

Presenting the Activity

Now that students have observed some features of a water cycle in a closed system, they can relate this process to what occurs in the natural environment. They will draw and label the processes in the water cycle as they compare what happened in the Ziploc bag with what happens in the water cycle in the natural environment.

ACTIVITY 2   Name: _____

                                    Date: _____

COMPARING WATER CYCLES

Inquiry Question: How does the water cycle in the closed system (Ziploc bag) relate to the water cycle in the natural environment?

Relate the process you observed in the closed system to the water cycle in nature. Describe what evidence you have that each process—evaporation, condensation, and accumulation—occurred in the Ziploc bag.

Process	*Evidence*
Evaporation	_____
Condensation	_____
Accumulation	_____

What does the process of precipitation look like in nature's water cycle?

_____

What did precipitation look like in the closed system?

_____

Draw the water cycle below as it occurred in the Ziploc bag. Label the area where each process occurred (evaporation, condensation, and accumulation).

A SALTY EXPERIENCE

Description of Activity

In this activity, students will investigate what happens when salt is added to the water in the closed system. You may want to have students set up this activity at the same time they are doing Activity 1 or within 1–2 days so that they will not have to wait another week for data. Or students may work in pairs; one student sets up a water cycle with plain water and the other student uses salt water.

By observing that the salt remains in the cup while the water goes through a cycle, students will infer that the water cycle distills the water, leaving impurities behind. Students are asked to make some connections to technology and society as they discuss the importance of this process for the earth's supply of fresh water.

Materials
• Graduated cylinders • Small plastic cups (3 oz.) • Large Ziploc bags • Tape • Metric measuring spoons • Salt

Presenting the Activity

If students have not done Activity 1, have them do this investigation at the same time or a day or so after they set up Activity 1. Ask the following question: "What would happen if salt were added to the water in the cup in a closed system? Predict and investigate as in Activity 1." If students have completed Activity 1, they will have some information about the way water cycles in a closed system. Now ask the following: "If salt were added to the water in the cup, what would happen to the salt during the water cycle? Would the salt stay with the water throughout the process or would it remain in the cup?" Students should make predictions and give a reason. Then, they will observe and collect data over the course of 4 days. They will observe and make inferences related to their observations.

In the writing prompt, students will research information about acid rain. They will consider how rain becomes acidic and how acid rain affects buildings, statues, and plants.

ACTIVITY 3

Name: _____

Date: _____

A SALTY EXPERIENCE

Inquiry Question: What would happen in the closed system if the water in the cup were salt water?

Repeat the procedures from Activity 1, but this time add 5 mL of salt to the 40 mL of water in the cup. Predict what will happen to the salt throughout the water cycle.

I predict _____

because _____.

Observe the conditions within the Ziploc bag each day for 4 days and record your observations on the chart below.

Day	Date/Time	Observations
1		
2		
3		
4		

Estimate how much water is at the bottom of the Ziploc bag.

_____

Is the water in the cup salty or plain? _____

Is the water on the bottom of the Ziploc bag salty or plain? _____

What can you conclude about the water cycle?

_____

_____

_____

_____

How is the supply of fresh water in the world dependent on the water cycle?

_____

_____

_____

_____

Based on what you learned in the activity, describe what the "ideal" quality of rain water would be.

_____

_____

_____

CRITERION-REFERENCED TEST:
WATER, WATER EVERYWHERE

Answer the following multiple-choice questions on the topic of water cycles.

1. In this diagram representing the water cycle, the process of evaporation is represented by which arrow?

   a. Arrow 1
   b. Arrow 2
   c. Arrow 3

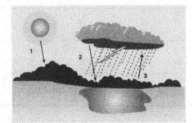

2. In the above diagram, the arrow representing the process of precipitation is
   a. Arrow 2
   b. Arrow 3
   c. Arrow 1

3. In the process of evaporation, liquid water is changed to
   a. frozen water
   b. precipitation
   c. clouds
   d. water vapor

4. In the process of condensation, water vapor is changed to
   a. water droplets
   b. a gas
   c. water vapor
   d. evaporated water

5. In the water cycle, water in the form of a gas changes back to its liquid form during the process of
   a. evaporation
   b. condensation
   c. precipitation
   d. accumulation

6. In the water cycle, water changes to water vapor through the process of
   a. evaporation
   b. precipitation
   c. adaptation
   d. condensation

7. Clouds and fog are made up of droplets of water that form on dust and other tiny particles of matter during the process of
   a. precipitation
   b. evaporation
   c. condensation
   d. accumulation

8. Rain, snow, and sleet are forms of
   a. precipitation
   b. condensation
   c. evaporation
   d. clouds

9. The source of energy for the water cycle is the
   a. atmosphere
   b. clouds
   c. sun
   d. water

10. All cycles are alike because they
    a. end at the same place
    b. have the same stages
    c. show how animals evolved
    d. are continuous

11. Water in the oceans would not be able to evaporate if it were not for the
    a. air
    b. salt
    c. sun
    d. icebergs

12. Water is placed in two shallow dishes. One dish is placed in a warm spot and one in a cool spot. What is likely to happen to the evaporation rate of the water in these two dishes?
    a. They will evaporate at the same rate.
    b. The one in the cool spot will evaporate first.
    c. The one in the warm spot will evaporate first.
    d. The water will not evaporate from these dishes.

13. The water cycle is a continuous process. Which of the following is NOT an example of a cycle?
    a. baking a cake
    b. the seasons
    c. rock formation and breakdown
    d. stages in the life of a frog

14. Life cycles and the water cycle are similar because
    a. they have no beginning and no end
    b. they both deal with the universe
    c. one is living and one is nonliving
    d. they have the same stages

WRITING PROMPT: ACID RAIN

Through these activities, you have learned that the water cycle distills water; that is, it separates water naturally from impurities. If this is so, how do you think rain water becomes acid rain?

Your task is to research acid rain and write about it. Your report should consist of several paragraphs that answer the following questions:

- How does rain become acid rain?
- What effect does acid rain have on buildings, statues, plants, etc.?
- What can be done to minimize or eliminate acid rain?
- Should people be concerned about acid rain and, if so, why?

You should use at least two different sources of information for your report. You may use written material, video resources, computer-accessed information, and/or human resources. Give a complete reference for each source used.

Discuss your findings with your classmates.

ANALYTIC RUBRIC: WATER, WATER EVERYWHERE

Indicators of Learning Relate to These Unifying Concepts and Processes

- Systems, Order, and Organization
- Evidence, Models, and Explanations
- Constancy, Change, and Measurement
- Evolution and Equilibrium
- Form and Function
- Mathematics, Literacy, and Thinking Skills

(Items in parentheses identify some of the concepts and skills assessed through the task.)

SCALE

	Complete	Almost	Not Yet	Comments
Activity 1: Water Cycle Observations **The student:** • Measured water accurately (Concept Understanding; Measurement) • Made a reasonable prediction • Gave a logical explanation for the prediction (Reasoning) • Recorded observations over four days • Estimated and recorded the amount of water • Made an inference about the location of the water in the closed system (Systems, Order, and Organization; Evidence, Models, and Explanation)				
Activity 2: Comparing Water Cycles **The student:** • Described evidence for each of three processes (Evidence; Explanation) • Described precipitation in the natural system (Explanation; Concept Understanding) • Described precipitation in the closed system (Inference; Explanation) • Drew a diagram of the water cycle and labeled processes in the water cycle (Systems, Order and Organization; Models and Explanation; Concept Understanding)				

	Complete	Almost	Not Yet	Comments
Activity 3: A Salty Experience **The student:** • Measured water and salt accurately • Made a reasonable prediction • Gave a logical explanation for the prediction • Recorded observations over four days • Estimated and recorded the amount of water at the bottom of the bag • Identified salty water in the cup and plain water in the bag (Evidence, Models, and Explanation) • Drew a conclusion about the water cycle (Concept Understanding) • Described the importance of the water cycle (Concept Application; S-T-S) • Described the "ideal" quality of rain water (Evidence, Models, and Explanations)				
Criteria for Acid Rain The student's report should: • Include information from at least two sources • Adequately answer each of the four research questions (Concept Understanding and Application; S-T-S) Teachers may add indicators for organizing and writing reports or mechanics of writing.				

Scoring Scale	
Complete	Student exhibited the indicator
Almost	Student showed some evidence that the indicator was exhibited, but something is incorrect or missing
Not Yet	Student did not show evidence of learning for that indicator

A Wholesome Partnership 9

Rationale

This performance activity assumes that students have experience using a balance and set of weights to find the mass of a variety of objects. They should understand how to calculate percentages and understand whole–part relationships. They should also have experience working with and constructing data tables. In this set of performance activities, both the terms *weight* and *mass* will be used. Intermediate level students are familiar with the term weight, but, as they are introduced to metric measurement, they should also be introduced to the term mass.

Students are asked to find the mass of the whole object (banana) as well as each of its two components: the white edible part and the yellow outer covering. Students should be able to explain any difference when the mass of the two parts does not equal the mass of the whole banana. It is important that students understand the fact that the mass of the whole object is always the same as the sum of the parts. Thus, any discrepancy should be attributed to such variables as the accuracy of the balance, types of weights used, precision of the procedure and researcher, data-recording techniques, and unexpected variables.

If discrepant results are found (the whole does not equal the sum of the parts), students should show persistence and a willingness to redo the activity for greater accuracy rather than be satisfied with results that do not show the expected outcome. They should be aware of variables that may affect their results.

In Activity 1, students will make estimates, record group data, find averages, and draw conclusions. Finally, they will be asked to develop a formula based on their measurements and defend the concept that the whole is equal to the sum of its parts.

National Science Education Standards

- Objects have many observable properties, including size, weight, shape, color, temperature, and the ability to react with other substances. (p. 127)
- Identify questions that can be answered through scientific investigations, design and conduct a scientific investigation, use appropriate tools and techniques to gather, analyze, and interpret data, develop descriptions, explanations, predictions, and models using evidence, and think critically and logically to make the relationships between evidence and explanations. (p. 145)

Project 2061 Statements from Benchmarks for Science Literacy

- No matter how parts of an object are assembled, the weight of the whole object is always the same as the sum of the parts; and when a thing is broken into parts, the parts have the same total weight as the original thing. (p. 77)
- Mathematical statements can be used to describe how one quantity changes when another changes. (p. 219)

Source: Reprinted with permission from **The National Science Education Standards** © **1997** by the National Academy of Sciences, Courtesy of the National Academies Press, Washington, D.C. Benchmarks for Science Literacy by permission of Oxford University Press, Inc.

In Activity 2, students will make an estimate about the percentage of the edible portion in very ripe, green, small, or large bananas or in some other variety of banana such as plantains or miniature bananas. Students will have an opportunity to make their own data tables, record data, draw conclusions, and compare findings with the results of Activity 1.

In an optional activity, students might be asked to research information about bananas such as the history, geographical origin, and nutritional value of bananas. Research groups might design instructional approaches for teaching what they learned to other groups. Creative presentations might include writing and performing a rap or song; designing and playing a game; writing and performing a play; or creating and displaying posters, charts, graphs, or other visuals to accompany a verbal presentation of information.

Teachers may wish to develop criteria and design an observation checklist for skills that will be assessed during this activity. Working with a group, researching information, staying on task, recording information, taking part in the performance, showing a positive attitude, and taking initiative are important skills for students to learn and use in this activity.

Source: The activities in this performance assessment are based on an activity called "The Big Banana Peel" from the book *Math + Science, A Solution,* developed by the AIMS Education Foundation, P.O. Box 8120, Fresno, CA 93747, http://www.aimsedu.org

SUM-THING FRUITFUL

Description of Activity

In this activity, students will first discuss their ideas on whole–part relationships. After the discussion, students will apply their ideas to bananas, estimating what percentage of the total mass of the banana they think the edible portion represents. Students may discuss ideas with each other, but they should record individual estimates. They will investigate the relationship by finding the mass of the whole object and masses of the edible and inedible parts. Students will record data, make graphs, and explain their findings.

Materials
• Balances • Mass sets • Paper towels • Bananas

Presenting the Activity

Ask students what they think about the concept that the total mass (weight) of the parts of an object is always the same as the mass (weight) of the whole object, no matter how the parts are assembled.

Encourage students to provide examples of how they know this concept.

After presenting bananas to students, ask them what percentage of the total they think the edible portion represents. Define "edible portion" as the white part that separates naturally from the outer yellow covering. Allow students to investigate the question, complete the data table, and record their observations. Then, challenge them to write a formula that represents the edible portion. Finally, have students collect and compare data from other groups or individuals, record their findings, and calculate the average percentage for all bananas.

Note: It may be necessary to review the procedures for finding mass and weight, and for controlling variables throughout the investigation. If students wish to put the edible portion of the banana on a paper towel while they take their measurements, the towel must also be used when finding the mass of the whole banana.

ACTIVITY 1   Name: _____

                                    Date: _____

SUM-THING FRUITFUL

Inquiry Question: What percentage of a banana is edible?

I think _____

because _____.

Find the mass of (1) the whole banana, (2) the edible portion, and (3) the outer yellow skin. Record the information on the data table below.

Object	Mass (in grams)	Ratio (edible/whole)	Decimal	Percent
Whole Banana	_____	_____	_____	_____
Edible Portion	_____	**Ratio** (skin/whole)	**Decimal**	**Percent**
Skin	_____	_____	_____	_____

What did you observe?

_____

_____

_____

_____

_____

_____

What can you say about the relationship between the parts of the banana and the whole banana?

_____

_____

_____

_____

What formula (using decimal or percent) would represent the edible portion of the banana?

Formula for the Edible Portion = _____

How do your results compare with the results of others? Collect data from five groups or individuals and record the data below.

Banana #	Mass of Whole Banana (g)	Mass Edible (g)	Mass of Skin (g)	Percent Edible (%)
1	_____	_____	_____	_____
2	_____	_____	_____	_____
3	_____	_____	_____	_____
4	_____	_____	_____	_____
5	_____	_____	_____	_____
Ours:	_____	_____	_____	_____

Find the average percentage edible for the *six* bananas.

Average Percentage = _____ %

ADDITIONAL LESSON IDEAS OF YOUR OWN

GOING BANANAS

Description of Activity

In Activity 2, students will reflect on what they found in Activity 1 and consider the whole-to-part relationships in bananas with different characteristics or bananas of different types.

Students will estimate the percentage for the new banana, find the mass of the whole and the parts, construct a data table, record data, and draw conclusions. Students will develop a formula for the new relationship and compare findings from Activities 1 and 2. They should explain how their investigations provide evidence for the concept introduced at the beginning of Activity 1.

Materials

- Balances
- Mass sets
- Paper towels
- Bananas with different characteristics from those used in Activity 1 (bananas that are very ripe, green, miniature, or extra large or a different variety)

Presenting the Activity

Ask students if they think the percentage of the edible portion of a banana is the same when the characteristics are different, such as when the banana is very ripe or when it is still green. Give students another banana with different characteristics to test. Directions are provided on the data sheet. Students are required to construct a data table. This may be done individually during the activity, or you may wish to have students design a data table as a group prior to conducting the investigation.

ACTIVITY 2

Name: _____

Date: _____

GOING BANANAS

Inquiry Question: Will the percentage of the edible part be the same for a banana with different characteristics from the one you previously investigated?

I think _____

because _____.

Based on what you learned in Activity 1, make a statement about the relationship between the whole fruit and the edible portion for the new banana.

Record the new banana's characteristics:

_____

List one way it is like the banana in Activity 1:

_____

List one way it is different from the banana in Activity 1:

_____

Make an estimate of the percentage of the edible portion of the new banana.
Percent Edible = _____

Find the mass of the whole banana, the edible portion, and the outer covering. Make a data table below and record the data. Calculate the percentage for the edible portion.

What conclusion can you draw from your data?

_____

Show a formula (percent edible) for this new banana:

_____

How do your findings compare with what you found in Activity 1?

_____

_____

_____

_____

CRITERION-REFERENCED TEST:
A WHOLESOME PARTNERSHIP

Answer the following multiple-choice questions regarding whole-to-part relationships.

1. Sue and her dad went to the store and bought a turkey for Thanksgiving dinner. The turkey had a mass of 5 kilograms. If the bones and inedible parts of the turkey had a mass of 2 kilograms, the mass of the edible portion was
   a. 3 kilograms
   b. 5 kilograms
   c. 2 kilograms
   d. 8 kilograms

2. Paul found that the mass of a large apple was 500 grams. He divided the apple into four parts and found that three had the following masses: 100 grams, 150 grams, and 125 grams. Paul estimated that the mass of the fourth piece should be
   a. 150 grams
   b. 125 grams
   c. more than 150 grams
   d. less than 125 grams

3. Suzy took a box with a glass bowl inside of it to the post office to be shipped. On the way to the post office, she dropped the box and heard a crash. The mass of the broken bowl in the box was
   a. the same as the mass of the box
   b. less than the mass of the bowl when it was whole
   c. more than the mass of the bowl when it was whole
   d. the same as the mass of the bowl when it was whole

4. Mr. Jones wanted to buy large boulders as decorations for his yard. At the landscaping store, he found three boulders, with weights of 30 lbs., 36 lbs., and 44 lbs. The total weight of the three boulders is
   a. 110 lbs.
   b. 100 lbs.
   c. 120 lbs.
   d. less than 100 lbs.

5. You found the weight of a large rock. Then you broke the rock apart into five smaller pieces. You weighed each piece and added the weights together. The total weight of the pieces was
   a. more than the weight of the large rock
   b. less than the weight of the large rock
   c. the same as the weight of the large rock
   d. similar to the weight of one piece

6. Franz has a bag of five apples. He found the weight of the whole bag of apples. Then he weighed each of the apples separately and the bag and added the weights together. What should Franz discover?
   a. The weight of five apples plus the bag is equal to the weight of the bag of apples.
   b. The weight of five apples is the same as the weight of the bag of apples.
   c. The individual apples weigh more than the bag of apples.
   d. The individual apples plus the bag weigh less than the bag of apples.

7. The candy store was having a sale on licorice at $2.00 per pound. Dinah wanted to buy some licorice but only had $1.00. She found Tony, who also had $1.00. Together, Dinah and Tony could buy how much licorice?
   a. one pound
   b. less than one pound
   c. more than one pound
   d. two pounds

8. John finds the mass of a plant to be 1 kilogram. He takes the plant apart and finds the mass of each individual part. Then he adds the masses of the parts and finds that they total
   a. less than 1 kilogram
   b. 1 kilogram
   c. more than 1 kilogram
   d. not even close to 1 kilogram

WRITING PROMPT: A WHOLESOME PARTNERSHIP

Jacques and Sylvia are working on Activity 1. They discover that the mass of the edible portion of their banana plus the mass of the outer covering is close but does NOT equal the mass of the whole banana. You have been asked to explain why results like this might occur.

What would you include in your explanation? Give your answer in a paragraph or two. In your explanation, be sure to provide the following:

- Two or more possible reasons for their findings
- An explanation of variables
- A mathematical explanation related to their results
- A plan for repeating the experiment for greater accuracy

ANALYTIC RUBRIC: A WHOLESOME PARTNERSHIP

Indicators of Learning Relate to These Unifying Concepts and Processes

- Systems, Order, and Organization
- Evidence, Models, and Explanations
- Constancy, Change, and Measurement
- Evolution and Equilibrium
- Form and Function
- Mathematics, Literacy, and Thinking Skills

(Items in parentheses identify some of the concepts and skills assessed through the task.)

SCALE

	Complete	Almost	Not Yet	Comments
Activity 1: Sum-Thing Fruitful **The student:** • Made a prediction based on logical reasoning • Measured the mass of the whole banana • Measured the mass of the edible portion • Measured the mass of the outer covering (Measurement; Recording Data) • Calculated the percent edible • Described an observation of data (Observation; Explanation) • Stated the whole–part relationship (Systems, Order, and Organization) • Developed a formula based on data • Completed a data table with information • Calculated the average percent				
Activity 2: Going Bananas **The student:** • Made a prediction • Gave a logical reason • Recorded characteristics of new banana (Observation; Explanation) • Identified a similarity (Comparing) • Identified a difference (Comparing) • Made an estimate of the percent edible • Found the mass of the new banana • Made a data table • Recorded data				

	Complete	Almost	Not Yet	Comments
• Calculated the percent of the edible portion • Drew a conclusion from data (Reasoning) • Showed a formula for new data (Concept Application and Understanding) • Compared findings to those in Activity 1 (Comparing)				
Criteria for A Wholesome Partnership The written explanation should include: • two or more possible reasons for the findings • an explanation of variables • a mathematical explanation related to their results (Evidence, Models, and Explanations; Concept Understanding) • a plan for repeating the experiment for greater accuracy (Concept Application; Reasoning)				

Scoring Scale	
Complete	Student exhibited the indicator
Almost	Student showed some evidence that the indicator was exhibited, but something is incorrect or missing
Not Yet	Student did not show evidence of learning for that indicator

A-W-L for One and One for A-W-L (Air-Water-Land)

10

Rationale

In this task, students will assume the role of environmental investigator for the Community Environmental Task Force. Students will choose to investigate air (Activity 1), water (Activity 2), or land (Activity 3) in their local environment for visible signs of pollution. This task should accompany a study of pollution in which the importance of pure air, clean water, and unpolluted land for the continued survival of living things is stressed.

Allow students to select one of the three activities for their investigation—Activity 1 to study air pollution, Activity 2 to study water pollution, or Activity 3 to study land pollution. The three activities are similar, and the data will vary only in terms of whether they refer to air, water, or land pollution. The type and amount of data resulting from an investigation will be determined by which activity is selected.

All students should do Activity 4, which asks them to prepare a short presentation using a visual aid to describe their investigation and findings. By

National Science Education Standards

- Abilities necessary to do scientific inquiry (pp. 122–123)
- Changes in environments can be natural or influenced by humans. Some changes are good, some are bad, and some are neither good nor bad. Pollution is a change in the environment that can influence the health, survival, or activities of organisms, including humans. (p. 140)
- Science cannot answer all questions and technology cannot solve all human problems or meet all human needs. (p. 169)

Project 2061 Statements from Benchmarks for Science Literacy

- Scientists do not pay much attention to claims about how something they know about works unless the claims are backed up with evidence that can be confirmed with a logical argument. (p. 11)
- Statistical predictions are typically better for how many of a group will experience something than for which members of the group will experience it—and better for how often something will happen than for exactly when. (p. 227)
- Students should keep records of their investigations and observations and not change the records later. (p. 286)
- Students should offer reasons for their findings and consider reasons suggested by others. (p. 286)

Source: Reprinted with permission from **The National Science Education Standards** © 1997 by the National Academy of Sciences, Courtesy of the National Academies Press, Washington, D.C. Benchmarks for Science Literacy by permission of Oxford University Press, Inc.

investigating signs of pollution in the immediate environment, students will realize that they can be personally affected by pollution.

AIRING OUR DIFFERENCES

Description of Activity

Students will investigate air samples to determine the types and amount of particles that they find.

Materials

- One transparency or sheet of plastic per student
- Petroleum jelly
- Three plastic baggies per student
- String
- Masking tape
- Hand lenses
- Hole punch

Presenting the Activity

In this task, students will assume the role of environmental investigator for the Community Environmental Task Force to study the quality of air in their community.

Students should think about and discuss what factors affect air quality. They should identify some of the types of materials they have seen going into the air. These might include airplane, bus, automobile, and motorcycle exhaust; smoke from factory, office, or home chimneys; smoke from burning leaves; excessive amounts of dust from fields; and pollen from plants.

Students should cut a sheet of plastic into four sections and punch a hole through the top of three of these sections. These sections will become their "particle detectors." Students should then thread a 10–12 cm piece of string through each hole and tie the strings so that the sections can be hung from or taped to objects. Next, they should smear petroleum jelly on one side of each particle detector. Students (or teachers) will select three sites to test. The particle detectors should be placed where there might be a lot of particulate matter being emitted, and they should be in places that are easily accessible. Students might select a place in or near their garage to collect particles from automobile exhaust, or a place that is dusty, or an area near the school to capture bus exhaust. The particle detectors should be placed where they will not be disturbed for at least 3 days.

After 3 days, students should collect the particle detectors and place each one in a separate plastic bag that is labeled with the location of the site. Students should examine the particle detectors with a hand lens to determine the types and amounts of particulate matter that were trapped. They should record their observations.

Before doing Activity 4, students should answer the questions on the student data sheet about the sources of particles, possible ways to reduce particles in the air, and the importance of clean air.

ACTIVITY **1**

Name: _____

Date: _____

AIRING OUR DIFFERENCES

Inquiry Question: What are the types and amounts of pollution found in the air in your community?

You have been asked to be a member of the Air Quality Team of the Community Environmental Task Force. The goal of the Task Force is to identify the types and amounts of pollution found in the air in your community. Your job is to sample the air in three places and determine the types and amounts of particles in the air. Following your investigation, you will prepare a report and give a presentation to community members who are concerned about the health hazards present in the local environment.

One type of pollution found in the air is particulate matter. Particles from dust, smoke, or chemicals may be present in air samples. Make a particle detector by taking a sheet of plastic and cutting it into four equal parts. Punch a hole at the top center of three pieces and tie a 10-centimeter piece of string through each hole so that the section can be hung from or taped to an object. Coat three pieces of plastic on one side with petroleum jelly. Now your particle detector is ready to be hung or taped in places where particles can be collected from the air.

Think about places in your community that might have a lot of particles in the air. Your task is to sample three places with high amounts of particulate matter in the air, if possible.

Identify the three places you will investigate and tell why you chose each place.

	Place	*Reason for Decision*
1.	_____	_____
2.	_____	_____
3.	_____	_____

Predict in which of the three places you expect to find the most particulate matter and explain why.

I predict _____

because _____.

Hang or tape your particle detectors at the three sites and wait 3 days before removing them. Remove the particle detectors and place each in a plastic bag that is labeled with the name of its location. Use a hand lens to observe the amount of particulate matter and the characteristics of the particles that were collected. Record each site's information on the following table.

Place	Amount of Particulate Matter (see scale below)	Texture/Shapes/Sizes
1. _____	_____	_____
2. _____	_____	_____
3. _____	_____	_____

Scale

LOW pollution: a few particles

MEDIUM pollution: concentrated spots or light film over all

HEAVY pollution: more spots or dark film over all

What do you think caused the particles to be in the air in the three places?

_____

_____

_____

_____

Can you suggest one way that the amount of particulate matter in the air can be reduced?

_____

_____

_____

_____

Why is it important to have clean air? What health problems are associated with polluted air?

_____

_____

_____

_____

WAT-ER WE FINDING IN OUR WATER?

Description of Activity

Students will take three samples of water from the immediate environment to investigate the presence of particles.

Materials

- Coffee filters
- Small metal soup cans in which about a dozen nail-size holes have been punched in the bottom
- Beakers or availability of a sink
- Hand lenses
- Collecting jars
- Labels for jars

Presenting the Activity

In this task, students will assume the role of environmental investigators for the Community Environmental Task Force to study the quality of the water in their community. Because there are various ways in which particles, chemicals, and unwanted materials can enter the water, students should think about some of the factors that affect water purity.

First, students will identify three areas in the community from which to gather water samples. They may select samples from ponds, streams, puddles, rain water, well water, tap water, river water, water from a golf course, or other sources. Students should collect approximately the same amount of water from each site and label the jars with the location of the site from which the water was taken.

A simple filter system can be constructed by fitting a coffee filter inside a soup can in which holes have been punched, or cutting filters to fit over the holes in the bottom of the can. Students will observe the color, smell, and other properties of the water and then pour the water through the can over a second container or sink. The filter paper will trap the particles that are in the water. Using a hand lens, students will observe the properties of the particles and record their observations on the data table.

Allow students to answer the reflective questions at the end of the activity before doing Activity 4.

ACTIVITY 2

Name: _____

Date: _____

WAT-ER WE FINDING IN OUR WATER?

Inquiry Question: What are the types and amounts of pollution found in the water in your community?

You have been asked to be a member of the Water Quality Team of the Community Environmental Task Force. The goal of the Task Force is to identify the types and amounts of pollution found in the water in your community. Your job is to sample water from three places and determine the types and amounts of pollutants in the water. Following your investigation, you will prepare a report and give a presentation to community members who are concerned about the health hazards present in the local environment.

One type of pollution often found in water is particulate matter. Particles and other pollutants may be present from litter, soil, chemicals, oil, or other materials. Dark color, foul smell, and the presence of particles are some indicators of pollution. Samples of water should be taken from three different places in the community. Think about places in your community where you might find polluted water. Try to find three water sources with high amounts of particulate matter in the water.

Identify the three places you decide to sample and tell why you chose those places.

Place	Reason for Decision
1. _____	_____
2. _____	_____
3. _____	_____

Predict which of the three places you expect to find the most particles in water and explain your prediction.

I predict _____

because _____.

Visit each of the three sites. At each location, fill a collecting jar with the same amount of water. Tape a label on each jar and record the location and the specific area within the location from which the water was taken (e.g., "The local pond—water from the shoreline about 4 cm under the surface").

After you have your samples, construct a filtering system to test for particles. Use a metal can that has holes punched in the bottom and carefully place a coffee filter into the can to cover the holes, keeping the opening as large as possible. If you prefer, you may cut a piece of filter paper just large enough to fit on the bottom of the can.

Observe the properties of the water in each jar (color, smell, presence of material, etc). Record the data on the chart. Over a sink or beaker, pour the contents of one collecting jar into the can and wait for the water to go through the coffee filter and out the bottom of the can. Remove the filter and label it for identification. Do this for the other two samples so that you have three filters from different areas to study.

Now, observe each of the filters. Estimate and record the amount of particulate matter in the sample using the scale that is given below the data table. Use a hand lens to observe the properties of the particles. Record their descriptions.

Sample/Location	Color/Smell	Amount	Description of Particles
1. _____	_____	_____	_____
2. _____	_____	_____	_____
3. _____	_____	_____	_____

Scale

LOW pollution: a few particles

MEDIUM pollution: concentrated spots or light film over all

HEAVY pollution: more spots or dark film over all

What do you think caused the particles to be in the water?

_____

_____

_____

_____

Can you suggest one way that the amount of particulate matter in water can be reduced?

_____

_____

_____

_____

Why is it important to have clean water?

_____

_____

_____

_____

EVERY LITTER BIT HURTS

Description of Activity

Students will investigate land areas in their community to determine the types and amounts of litter that they find.

Materials

- Large notebooks
- Pencils

Presenting the Activity

In this task, students will assume the role of environmental investigators for the Community Environmental Task Force. As investigators, they have been asked to study the problem of litter in their community.

Students should think about and discuss what factors determine the amount and types of unwanted litter that is found on the land. Students should select three sites where they will investigate litter. These should be safe places to visit or be places that they can go accompanied by a parent or other adult (i.e., a local park, school grounds, home yard, areas near home or school, areas along streets or near shopping malls).

Students will examine each of the areas and look for litter. They will describe the types and record the amounts of litter found at each site. Using the scale given, they will gauge the amount of litter found as a low, medium, or high level. Following their investigations, students should answer the questions concerning sources of litter, possible ways to reduce litter, and the importance of litter-free land before doing Activity 4.

Name: _____

Date: _____

EVERY LITTER BIT HURTS

Inquiry Question: What are the types and amounts of litter found in your community?

You have been asked to be a member of the environmental investigators of the Community Environmental Task Force. The goal of the Task Force is to identify the types and amounts of pollution found on the land in your community. Your job is to sample three places to determine the types and amounts of litter on the land. Following your investigation, you will prepare a report and give a presentation to community members who are concerned about the health hazards present in the local environment.

One type of pollution that is found on the ground is litter. Litter is usually defined as paper, plastic, glass, wood, and other types of waste or trash that are not a natural part of the environment. Litter may be brought to an area by animals, wind, or humans. Think about places in your environment that might have a lot of litter. Your task is to visit three land areas in your local environment, make observations, and record the types and amount of litter that are present.

Identify the three places you decide to investigate and tell why you chose those places.

Place	*Reason for Decision*
1. _____	_____
2. _____	_____
3. _____	_____

Predict which of the three places you expect to find the most litter and explain your prediction.

I predict _____

because _____.

For each area, record the amount of litter (use the scale to help describe the amount) and the types of litter found. Use the back of the page if you need additional space.

Location	Amount of Litter	Types of Litter
1. _____	_____	_____
2. _____	_____	_____
3. _____	_____	_____

Scale

LOW pollution: a few pieces of litter

MEDIUM pollution: litter obvious, several different types

HEAVY pollution: lots of litter, various types, area looks cluttered/messy/overrun

What do you think caused the litter to be in the area?

_____

_____

Can you suggest one way that the amount of litter can be reduced?

_____

_____

Why is it important to have a clean, litter-free environment?

_____

_____

ADDITIONAL LESSON IDEAS OF YOUR OWN

PREPARING AND GIVING AN ENVIRONMENTAL TASK FORCE REPORT

Description of Activity

As members of the Community Environmental Task Force, students will prepare and give a presentation of their findings to the community. You will decide how much time students will have for their presentations. Each student should create a poster, model, collage, or other visual to help describe his or her findings and present data; make inferences about causes of pollution; and offer suggestions for reducing the amount of air, water, or land pollution in the environment.

Materials

- Poster board, crayons, markers, etc., for making visual aids

Presenting the Activity

As members of an Environmental Task Force, students have a responsibility to share their findings, inferences, and suggestions for improving the quality of the environment with members of the community. They should prepare a presentation using the list of criteria on the following page to guide their work.

Note: Please add information for students such as time guidelines for the presentation, use of materials, and other guidance appropriate for your class.

ACTIVITY 4   Name: _____

                                            Date: _____

PREPARING AND GIVING AN ENVIRONMENTAL TASK FORCE REPORT

Inquiry Question: How do members of an Environmental Task Force share the results of their investigations with other members of the community?

As a member of an Environmental Task Force, you have a responsibility to share your findings with citizens who are concerned with the health and welfare of their community. You are to prepare and give a presentation to the community on the findings of your investigation of air, water, or land pollution.

Prepare a presentation that includes the following:

1. A complete description of your project, identifying the type of pollution you studied and the three sites you chose to study
   a, One or more visual aids (posters, pictures, models, graphs, drawings, etc.) to display data or relevant information
   b. Findings of your investigation
   c. Inferences about what might have caused the pollution
   d. Evaluation of the sites, your procedures, and your data

2. Sites for investigation: Were they good places? Did you find what you were looking for in these places? Were there other places you wish you had studied?

3. Procedures: Were you careful with the materials you used? Did you follow directions? Did you study the samples carefully enough to get useful data? Were there other things you could have done to get better results?

4. Data: Did you get enough data to make inferences or draw conclusions? Why or why not? If you were to do this investigation again, do you think you would get the same or different results? Explain.
   a. Suggestions for further investigations
   b. Suggestions to the community for improving the quality of their environment

ANALYTIC RUBRIC: A-W-L FOR ONE AND ONE FOR A-W-L

Indicators of Learning Relate to These Unifying Concepts and Processes

- Systems, Order, and Organization
- Evidence, Models, and Explanations
- Constancy, Change, and Measurement
- Evolution and Equilibrium
- Form and Function
- Mathematics, Literacy, and Thinking Skills

(Items in parentheses identify some of the concepts and skills assessed through the task.)

SCALE

	Complete	Almost	Not Yet	Comments
Activities 1, 2, and 3: Airing Our Differences, Wat-er We Finding in Our Water?, and Every Litter Bit Hurts **The student:** • Identified three places and gave reasons for the decisions (Explanation) • Made a prediction • Gave a logical reason for the prediction • Identified the location, amount, and description of pollutants for each of three sites selected (Evidence, Models, and Explanations) • Inferred causes for the pollution • Suggested a way to reduce the pollution (Concept Understanding and Application; S-T-S) • Described the importance of a pollution-free environment (Concept Understanding; Explanation)				
Activity 4: Preparing and Giving an Environmental Task Force Report **The student:** • Gave a complete description of the project • Created and used a visual to enhance the presentation (Representation)				

	Complete	Almost	Not Yet	Comments
• Described findings (Explanation) • Made inferences about what caused the pollution (Systems, Order, and Organization) • Evaluated the sites • Evaluated procedures used to investigate • Evaluated results of investigation (Reasoning) • Made suggestions for further investigations • Made suggestions for improving the quality of the environment (Concept Application; S-T-S)				

Scoring Scale	
Complete	Student exhibited the indicator
Almost	Student showed some evidence that the indicator was exhibited, but something is incorrect or missing
Not Yet	Student did not show evidence of learning for that indicator

The Mysterious Package 11

Rationale

This task consists of one substantial activity that gives students an opportunity to investigate the diet of a barn owl firsthand. Students should have prior knowledge of the ways organisms are categorized by the function they serve in an ecosystem: producers, consumers, and decomposers. They should also have some knowledge of food chains and food webs.

Barn owls are expert hunters with specially designed eyes, ears, claws, and wings used to catch prey. Owls often swallow their prey whole. They cannot digest bones, fur, and feathers, so these are clumped together and regurgitated as "pellets." When owl pellets are carefully pulled apart, they yield bones and skulls of rodents or birds. When the activity is presented in a positive manner, it is an excellent, high-interest study of a food chain. Students are fascinated by what they find in the small package!

Ultimately, each student should find evidence of at least one animal that can be identified from pictures on charts showing the various prey. As an extension activity, students can

National Science Education Standards

- All animals depend on plants. Some animals eat plants for food. Other animals eat animals that eat the plants. (p 129)
- Populations of organisms can be categorized by the function they serve in an ecosystem. All animals, including humans, are consumers, which obtain food by eating other organisms. (pp. 137–138)

Project 2061 Statements from
Benchmarks for Science Literacy

- A great variety of living things can be sorted into groups in many ways using various features to decide which things belong to which group. (p. 103)
- Almost all kinds of animals' food can be traced back to plants. (p. 119)
- Some source of "energy" is needed for all organisms to stay alive and grow. (p. 119)
- Two types of organisms may interact with one another in several ways: They may be in a producer/consumer, predator/prey, or parasite/host relationship. (p. 117)
- Tables and graphs can show how values of one quantity are related to values of another. (p. 218)
- The graphic display of numbers may help to show patterns such as trends, varying rates of change, gaps, or clusters. (p. 224)

Source: Reprinted with permission from **The National Science Education Standards** © 1997 by the National Academy of Sciences, Courtesy of the National Academies Press, Washington, D.C.
Benchmarks for Science Literacy by permission of Oxford University Press, Inc.

reconstruct the skeletons they find and display them. They should draw food chains by researching the types of food that are eaten by the prey. Most prey are herbivores, requiring students to trace the food chain back to plants and to the ultimate source of energy for all of life on earth—the sun.

LARGE THINGS COME IN SMALL PACKAGES

Description of Activity

In this task, students will investigate owl pellets. Owl pellets provide firsthand information about food chains and skeletal anatomy. **Note of caution:** For the safety of your students, use *only* owl pellets that have been heat sterilized, such as those available commercially. The pellets will reveal an assortment of animal bones and skulls that students can classify by matching them to pictures on identification charts. The best part of this activity is that each pellet is different, adding elements of variation and surprise.

Have students work individually or in pairs to measure and examine the owl pellets before dissecting them. They should observe, sort, and record findings and make individual bar graphs and a class graph to show the types and number of prey found. Be sure to allow plenty of time for finding, sorting, reconstructing, and sharing information. Students may need several sessions to complete this activity.

As they describe and explain their findings, students should show an understanding of food chains and food webs and the transfer of energy. In the writing prompt, students will apply what they learned to solve a problem.

Materials

- Owl pellets and prey charts (pellets and charts can be ordered from Pellets, Inc., P.O. Box 5484, Bellingham, WA 98227-5484; 1-888-466-OWLS; http://pelletsinc.com)
- Tweezers
- Probes (or toothpicks)
- Paper towels
- Egg cartons (optional)
- Notebooks for data and drawings

Presenting the Activity

Scientists (zoologists and field biologists) who are interested in learning about animals, their habitats, food, behaviors, and so forth are informed by evidence they can experience firsthand. Explain to students that the study of owl pellets provides immediate information on the diet of the barn owl. Emphasize that owl pellets are extraordinary because they provide a great deal of information about the barn owl's diet and the population of animals in the areas where the pellets are found. (Commercially purchased pellets have been sprayed and wrapped, but otherwise they are very much like they were when collected from the barn where the owls roosted.) By dissecting owl pellets, students can infer the total number and types of prey an owl consumed. If students are given a positive introduction to the investigation, they will be eager to investigate the pellets.

Students should measure and describe the pellets before dissecting them. Provide sorting sheets and charts to students to help them identify bones and skulls. Egg cartons may be used to sort bones. Ask students to reconstruct the skeleton of one or more of the owl's prey as an extension activity. Students may need to work in pairs or triads in order to have all the bones they need for one good specimen.

There are a number of social issues related to the study of owls. The habitats of owls are diminishing in some parts of the United States. Harsh weather and shortages of food pose serious threats to the survival of some species of owls. Students should conduct research to better understand the problems and issues related to the survival of these animals.

ACTIVITY **1**

Name: _____

Date: _____

LARGE THINGS COME IN SMALL PACKAGES

Inquiry Question: How do owl pellets help us understand food chains and food webs?

Measure and record the length of your owl pellet.

_____

Measure and record the circumference of your owl pellet.

_____

Take the wrapper off and describe your owl pellet.

_____

Dissecting your owl pellet: Use a tweezer to *carefully* probe and separate the bones from the fur, feathers, and other material in your pellet. Clean the bones and/or skulls and sort them by type using a sorting sheet, an egg carton, or labeled sections on a piece of paper.

On the chart, sketch each type of bone you found and record how many of each type were in the pellet.

Type and Sketch of Bones or Skulls	Number

Make a bar graph of the types and number of bones in your owl pellet.

If an owl consumes two prey each day, how many will it eat in a year? _____

Consider your findings as well as those of others in the class. What can you infer about the population of prey in the area where the pellets came from?

_____

_____

_____

_____

Draw a food chain with the barn owl as the top predator; label all the components. Be sure to include producers and consumers as well as the source of energy for the food chain.

CRITERION-REFERENCED TEST:
THE MYSTERIOUS PACKAGE

Answer the following multiple-choice questions related to the topic of food chains and food webs.

Consider the food chain:

PLANTS ⟶ GRASSHOPPER ⟶ TOAD ⟶ RACCOON

1. Which organism is the producer?
    a. plants
    b. grasshopper
    c. toad
    d. raccoon

2. Animals that eat plants are called
    a. carnivores
    b. herbivores
    c. top predators
    d. producers

3. Plants get the energy to produce food from
    a. soil
    b. photosynthesis
    c. animals
    d. the sun

4. Organisms that are able to make their own food are classified as
    a. producers
    b. consumers
    c. decomposers
    d. herbivores

5. Consider the food chain:

CARROT ⟶ RABBIT ⟶ HUMAN

    The herbivore is the
    a. human
    b. carrot
    c. rabbit
    d. top predator

6. Which of the following would be the lowest consumer on the food chain in #5?
    a. herbivore
    b. carnivore
    c. top predator
    d. plant

7. A food chain would not exist without
    a. sand
    b. predators
    c. carnivores
    d. sunlight

8. Green plants are very important in the food chain because they are the
   a. producers
   b. consumers
   c. decomposers
   d. herbivores

9. The energy for a food chain is provided by
   a. plants
   b. the sun
   c. herbivores
   d. carnivores

WRITING PROMPT: THE CHANGING ENVIRONMENT

A community of owls lives in an old red barn near a small but growing town in the state of Washington. The owls have been there for over 15 years, but it seems that food is harder to find every year. Numerous subdivisions and strip malls have been built over the years, bringing people closer and closer to the barn. You have learned that the barn may be taken down next year and the land may be sold for a housing development.

The Owl Preservation Society has asked for your help in designing a half-page ad to be placed in the local paper to inform readers of the contributions that owls make to the community and the importance of preserving habitats for them. The ad should include one or more drawings as well as a written message. The Preservation Society hopes that the citizens of the area will realize the importance of owls to the environment and will understand how the loss of the owls might affect the area.

Use the information you found in your "mysterious package" along with what you know about owls to design the half-page ad. Use the library or the Internet to conduct research on the barn owl to make your ad as informative and convincing as possible. The owls are depending on you to help preserve their habitat.

ANALYTIC RUBRIC: THE MYSTERIOUS PACKAGE

Indicators of Learning Related to Unifying Concepts and Processes

- Systems, Order, and Organization
- Evidence, Models, and Explanations
- Constancy, Change, and Measurement
- Evolution and Equilibrium
- Form and Function
- Mathematics, Literacy, and Thinking Skills

(Items in parentheses identify some of the concepts and skills assessed through the task.)

SCALE

	Complete	Almost	Not Yet	Comments
Activity: Large Things Come in Small Packages **The student:** - Measured the length of the owl pellet - Measured the circumference of the owl pellet - Described the pellet (Observation and Explanation) - Sorted the bones by type (Classification) - Sketched each type (Representation). Recorded number of bones of each type (Record Data) - Made a bar graph for data (Graphing) - Calculated number of prey accurately - Made an inference about the population of prey - Drew a food chain including owl and prey (Systems, Order, and Organization; Evidence, Models, and Explanation) - Included food for prey (Form and Function; Concept Understanding) - Showed the sun as the source of energy (Concept Understanding) - Labeled all components of the food chain correctly (Concept Understanding)				

	Complete	Almost	Not Yet	Comments
Criteria for The Changing Environment The half-page ad: • Included one or more drawings (Representation) • Included a written message related to problem (Explanation) • Used information learned in the activity (Concept Application) • Showed an understanding of food chain (Systems, Order, and Organization; Concept Understanding) • Made a convincing argument for the importance of the survival of owls and the need for habitat preservation (Systems, Order, and Organization; Concept Application)				

Scoring Scale	
Complete	Student exhibited the indicator
Almost	Student showed some evidence that the indicator was exhibited, but something is incorrect or missing
Not Yet	Student did not show evidence of learning for that indicator

Up, Up, and Away

12

Rationale

Activities and experiences that allow students to ask questions and investigate using equipment and materials are critical to the primary- and middle-grade science program; they help build a base for concept development, aid in the development of process and thinking skills, and provide a rich environment for the development of scientific literacy.

In addition to developing action plans for answering operational questions, students should understand how to set up and conduct a controlled experiment in which they state a hypothesis based on an inquiry question and investigate the effect of one variable on another.

Prior to this task, students should have experience identifying variables and conducting experiments in which they test the effects of one variable on another. For example, early elementary students might test effects of light, amount of water, and use of fertilizers on plant growth. Intermediate-grade students might test the distance a toy car will travel on ramps of various heights or with different types of surfaces.

In Activity 1, students will make a simple kite and test the effects of two variables on the performance of the kite. The variables include (a) the point of attachment of the string to the kite, and (b) the length of the kite's tail.

National Science Education Standards

- Identify questions; design and conduct investigations; use tools and techniques to gather, analyze, and interpret data; develop descriptions, explanations, predictions, and models using evidence; think critically and logically to make the relationships between evidence and explanations; recognize and analyze alternative explanations and predictions; communicate scientific procedures and explanations; use mathematics in all aspects of scientific inquiry. (pp. 145 and 148)

Project 2061 Statements from Benchmarks for Science Literacy

- Scientists differ greatly in what phenomena they study and how they go about their work. Although there is no fixed set of steps that all scientists follow, scientific investigations usually involve the collection of relevant evidence, the use of logical reasoning, and the application of imagination in devising hypotheses and explanations to make sense of the collected evidence. (p. 12)
- If more than one variable changes at the same time in an experiment, the outcome of the experiment may not be clearly attributable to any one of the variables. It may not always be possible to prevent outside variables from influencing the outcome of an investigation, but collaboration among investigators can often lead to research designs that are able to deal with such situations. (p. 12)
- Thinking about things as systems means looking for how every part relates to others. (p. 265)

Source: Reprinted with permission from **The National Science Education Standards** © **1997** by the National Academy of Sciences, courtesy of the National Academies Press, Washington, D.C. Benchmarks for Science Literacy by permission of Oxford University Press, Inc.

Activity 2 allows students to make decisions about the type of tail (material used or design) that enables the kite to fly best. Students will have the opportunity to be creative and imaginative in their decisions.

The rubric for this assessment deals with the indicators of learning related to conducting a controlled experiment.

GO FLY A KITE

Description of Activity

In this activity, all students will build simple kites to the same specifications, including the type and length of kite tails. Next, students will determine the point at which the string should be attached for the best performance of the kite. Three possible points of attachment will be offered: a front point (6 cm from front of the kite), a midpoint (7 cm from front of the kite), or a rear point (8 cm from front of the kite). Although the kites are identical, the results of the experiments may be different. Uncontrolled variables such as slight differences in the process of making the kites and conditions for flying the kites (wind conditions, position of students in relation to the wind, buildings and obstructions, etc.) may account for differences in results.

Students should keep a detailed account of what they test, where they test, and the results of each trial. If students are allowed to give verbal explanations of their experiences, teachers can learn a great deal about their understanding of the experimental process. Teachers may determine that skills and dispositions like reasoning skills, record keeping, communication, and respect for evidence may be assessed in this performance activity. Teachers may develop a checklist of skills and dispositions to observe as students carry out the investigations.

Materials

- 8.5 in. x 11 in. sheet of paper per person (colored paper, if possible)
- 1 plastic straw per person
- Transparent tape
- Metric rulers
- Pencils
- Kite string
- Paper punch
- Kite directions

Presenting the Activity

Explain to students that the local park district is offering a summer program for young children that will teach them about kite aerodynamics and provide an opportunity to make and fly simple kites. They have asked students to investigate the conditions under which the kites will fly best. The students will focus on the position at which the string is attached to the kite as well as the length and material of the kite's tail.

In this first activity, students will investigate the position at which the string should be attached for the best performance of the kite. Students will assess performance based on the following possible ways the kite may fly:

a. "Dances wildly" out of control or dives around in an unstable manner high in the sky with a loose string
b. "Pulls and dives" aimlessly, close to the ground
c. "Floats smoothly," looking light and stable in the air

Letter "c" is the most desirable kite performance. Tell students they must determine if the string should be attached at the front point (6 cm from the front of the kite), at the midpoint (7 cm from the front of the kite), or at the rear point (8 cm from the front of the kite) for the best kite performance.

Whether or not students have made or flown kites, they have likely seen a kite in the sky. Discuss what a kite looks like when it is flying smoothly, high in the sky. Ask students to think about what conditions might affect the way a kite flies. Make a list of conditions.

Have students make the straw kite according to the kite directions provided. When the kites are completed, students will test each attachment point. Ask them which of the three attachment points made the kite fly smoothly.

Students may conclude that the kite did not fly well at any attachment point with the design they used. Explain to students that the tail design may make a difference in the way the kites fly. In Activity 2, they will experiment with various tail designs and lengths.

Kite Directions

To make a kite, you need one sheet of 8.5 x 11-inch paper, one plastic straw, a pencil, a paper punch, metric ruler, and tape.

1. Fold the sheet of paper in half the long way. Each half sheet of paper will be 5.5 inches wide.

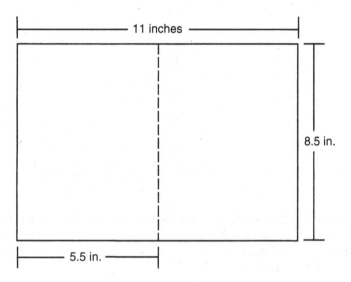

2. Measure 3 cm from the folded edge and draw a line from top to bottom.

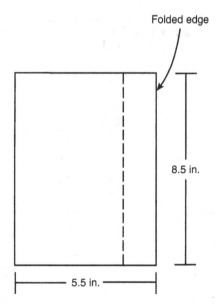

3. Crease the paper at the line and fold it over. The folded section is called the keel of the kite. Place the keel face down and open the paper, leaving the folded part (keel) underneath. Tape along the center line.

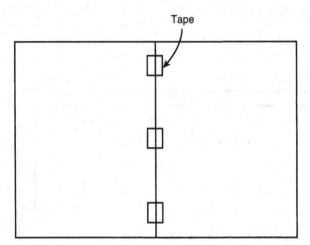

4. With the open paper facing you and the keel underneath, measure 2 cm from the top of the kite and draw a line across the kite. Tape a plastic straw along the line and secure it with a piece of tape at both ends and one in the middle. Turn the kite over and straighten out the keel.

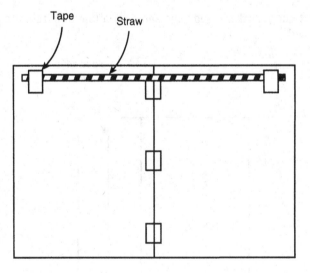

5. Measure 6 cm from the top of the kite. Put a mark on the keel about 1.5 cm from the edge at the 6 cm (from top) spot. Then measure another centimeter below that and put a mark at 7 cm, and put another mark at 8 cm. Reinforce the marks by placing tape over them and punch holes at the 6 cm, 7 cm, and 8 cm marks. These three holes will be the attachment points.

6 cm = front point

7 cm = midpoint

8 cm = rear point

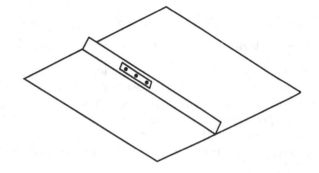

6. Make a tail for the kite from paper, plastic, tissue paper, etc. (If all kites are to be the same to start with, the students should decide on a design and length for the kite tail.)

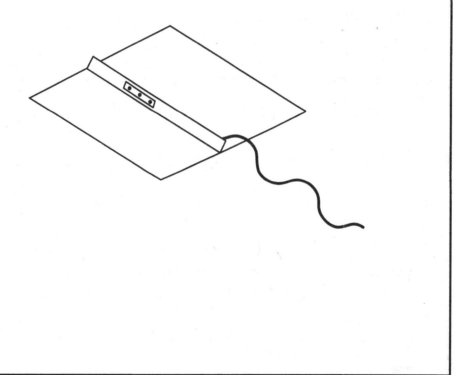

ACTIVITY 1

Name: _____

Date: _____

GO FLY A KITE

Inquiry Question: Which of the three points of attachment will enable your kite to fly with a smooth, floating movement?

You have been asked to help the local park district officials investigate kite aerodynamics. They have developed a new design for a kite to use with young children in a summer program, but they need your help to determine the best place on the keel of the kite to attach the string and the best type of tail to use so that the kite will fly smoothly. Make a kite according to the kite directions provided to you. As a class, decide on a design (type of material and length) for the tail. (In this investigation, all students should use the same tail design.)

There are three possible attachment points for the string: 6 cm, 7 cm, or 8 cm from the top of the kite. You are to test each of the three points of attachment to determine at which point the kite will fly the best. That is, select the point at which the kite will fly smoothly, looking light and stable in the air. It should not dance wildly out of control or dive in an unstable manner with a loose string. It also should not pull and dive aimlessly, close to the ground.

Make a prediction about which point will enable the kite to fly best. This is called stating a hypothesis:

The kite will fly best at the _____ point

because _____.

Design a plan for answering the research question. Provide a step-by-step explanation of your plan.

Carry out your investigation. Be sure to keep accurate records of data collected.

Description of the Kite's Tail

Material used: _____

Length in cm: _____   Width in cm: _____

Attachment Point	Description of Kite's Flight
Front point (6 cm)	_____
Midpoint (7 cm)	_____
Rear point (8 cm)	_____

Based on what you learned through your experimentation, what can you conclude about the best attachment point for your kite?

_____

_____

SMOOTH TAILING

Description of Activity

In this activity, students will investigate different tail designs or lengths to determine what effect they have on the performance of the kite. Students will decide individually or with a partner what tail designs or lengths they will compare to their first design. For each new design or tail length, the students should test the front, mid, and rear attachment points to determine which will give the best kite performance.

This performance task requires students to set up a minimum of two new experiments. For each experiment, students should write a description of their new design, document their testing procedures, and keep accurate records of the results. The structure of Activity 1 should prepare them to handle Activity 2, as a similar experimental design can be used.

Because the kite tails are different for each student or pair, results will be different. Students will test their kites and record their results, environmental conditions, types of materials and lengths used for kite tails, and cause-and-effect relationships. Students should realize that the more types and lengths of tails are tested by all students, the more data they will have upon which to base their conclusions.

Materials

- Straw kite from Activity 1
- Kite string
- Various types of paper strips such as newspaper, computer paper, or tissue paper
- Strips of plastic from garbage bags or bread bags
- Scissors
- Tape
- Metric rulers or meter sticks

Presenting the Activity

Explain to students that, although their kites may have flown well in Activity 1, their kites may perform even better with different kite tails. Additional investigations testing a variety of tail designs may provide the best overall design for the kite that the park district should use. Kite tails can be made of paper, plastic, tissue paper, or other material, and they can vary in length.

Students are asked to research at least two different tail designs to determine if their kites will fly better than they did in Activity 1. Students may change the material used for the tail, the length, or both. For each tail design, students should again test the three attachment points (6 cm, 7 cm, and 8 cm) to determine which attachment point will give the best (smoothest) flight pattern.

Activity 2

Name: _____

Date: _____

SMOOTH TAILING

Inquiry Question: Which type of kite tail will provide the best flight performance?

For each new design, test the three attachment points (6, 7, and 8 cm) on the kite as you did in the first activity.

New Tail Design I

Describe a new tail design. Tell how it is different from the kite tail in Activity 1.

New design: _____

How it differs from the tail in Activity 1: _____

_____

_____

Describe your plan for testing the new kite tail design.

_____

_____

_____

_____

_____

Conduct the experiment and record observations and other data.

_____

_____

_____

_____

Draw a conclusion based on the data you collected.

_____

_____

_____

_____

New Tail Design II

Describe the new tail design. Tell how it is different from the kite tail in Activity 1.

New design: _____

How it differs from the tail in Activity 1: _____

_____

Describe your plan for testing the new kite tail design.

_____

_____

_____

_____

Conduct the experiment and record observations and other data.

_____

_____

_____

_____

Draw a conclusion based on the data you collected.

_____

_____

_____

_____

WRITING PROMPT: KITE DESIGN

A reporter from the local newspaper has asked you to write a feature article describing the processes you used and the results of your experiments testing the *variables*—that is, the various conditions you tested—that influence kite flight. She is interested in sharing the results of your work with the community. The park district welcomes the publicity to promote their summer program.

Please help by writing an article that describes your research. Include a final conclusion about the best kite design after considering the data from your three experiments as well as the data and conclusions from other members of your class.

ANALYTIC RUBRIC: UP, UP, AND AWAY

Indicators of Learning Relate to These Unifying Concepts and Processes

- Systems, Order, and Organization
- Evidence, Models, and Explanations
- Constancy, Change, and Measurement
- Evolution and Equilibrium
- Form and Function
- Mathematics, Literacy, and Thinking Skills

(Items in parentheses identify some of the concepts and skills assessed through the task.)

SCALE

	Complete	Almost	Not Yet	Comments
Activity 1: Go Fly a Kite **The student:** • Made a prediction based on reasoning • Gave a description of the tail of the kite (Explanation) • Provided measurements of the kite tail in cm • Developed a plan for testing kite flight (Models and Explanation) (Teachers may determine the criteria for the plan such as: • Identified dependent and independent variables • Explained a logical experiment • Explained a sequence of steps • Designed a data table • Recorded attachment points and flight performances • Drew a conclusion based on data)				
Activity 2: Smooth Tailing **The student:** • Described the new tail design (Form and Function; Explanation) • Described differences in design • Made a prediction based on reasoning (Similarities and Differences) • Described a plan (additional criteria may be added) (Systems, Order, and Organization) • Designed a data table • Recorded data of flight performance for attachment points • Drew a conclusion based on data				

	Complete	Almost	Not Yet	Comments
Suggested Criteria for Kite Design The article should include an accurate description of the procedures and the data. Indicators of language arts may also be assessed, such as the mechanics of writing a paragraph.				

Scoring Scale	
Complete	Student exhibited the indicator
Almost	Student showed some evidence that the indicator was exhibited, but something is incorrect or missing
Not Yet	Student did not show evidence of learning for that indicator

Picture Perfect Professions

13

Rationale

Throughout their study of science topics, students will become increasingly aware of the vast number of occupations that are possible in the fields of science and technology. Students should read about occupations in science and technology and interview adults and professionals in related fields whenever possible.

It is not unusual for aspiring architects, artists, illustrators, writers, and others in arts-related fields to develop portfolios of their work to take to job interviews. Besides observing personal traits and perspectives during an interview, a prospective employer can view the work of the candidate and get a first-hand understanding of the artist's style and ability. Based on this idea, students are asked to consider how other professional candidates "sell themselves" to prospective employers.

National Science Education Standards

- Many people choose science as a career and devote their entire lives to studying it. Many people derive great pleasure from doing science. (p. 141)
- Women and men of various social and ethnic backgrounds – and with diverse interests, talents, qualities, and motivations – engage in the activities of science, engineering, and related fields such as the health professions. Some scientists work in teams, and some work alone, but all communicate extensively with others. (p. 170)
- Science requires different abilities, depending on such factors as the field of study and type of inquiry. Science is very much a human endeavor, and the work of science relies on the basic human qualities, such as reasoning, insight, energy, skill, and creativity—as well as on scientific habits of mind, such as intellectual honesty, tolerance of ambiguity, skepticism, and openness to new ideas. (p. 170)

Project 2061 Statements from
Benchmarks for Science Literacy

- People can learn about others from direct experience, from the mass communications media, and from listening to other people talk about their work and their lives. (p. 154)
- Each culture has distinctive patterns of behavior, usually practiced by most of the people who grow up in it. (p. 155)
- As students begin to think about their own possible occupations, they should be introduced to the range of careers that involve technology and science, including engineering, architecture, and industrial design. Through projects, readings, field trips, and interviews, students can begin to develop a sense of the great variety of occupations related to technology and to science and what preparation they require. (p. 46)

Source: Reprinted with permission from **The National Science Education Standards** © 1997 by the National Academy of Sciences, Courtesy of the National Academies Press, Washington, D.C.
Benchmarks for Science Literacy by permission of Oxford University Press, Inc.

This performance task is best done at the end of a school year so that students will be able to apply information they have collected about careers in science and technology throughout the year. The task challenges students to design a creative job interview to show their understanding of concepts, skills, and habits of mind related to a profession in science or technology.

Activity 1 is a basic research activity. Students are asked to find information related to a career in one of the fields of science or technology they have studied. They will create a photo journal by selecting four significant features of their chosen career and draw pictures that represent each feature. They might also choose to portray themselves as a professional involved in one or more of the significant aspects of the profession.

In Activity 2, students plan a job interview that relates to their chosen career area in science or technology. This task allows students to demonstrate their understanding of the concepts, skills, and habits of mind that are a natural part of the science- or technology-related profession. There is a powerful S-T-S (science-technology-society) connection in this two-part task, as students will be analyzing and describing the profession in terms of the applications of science to technology and society.

Prior to Activity 1, you may want students to engage in a discussion of the qualities and characteristics that employers seek in candidates for positions. They may offer examples related to the professionals they know and interact with such as teachers, coaches, doctors and nurses, and business professionals. Students should demonstrate some awareness of the need for an adequate knowledge base, job-related skills, and attitudes and values critical to effective job performance. Habits of mind such as honesty, persistence, logic and reasoning, cooperation, and responsibility should also be considered.

A PHOTO JOURNAL OF MY PROFESSION

Description of Activity

In this activity, students will research a career related to one of the areas of science that they have studied during the year. After conducting their research, they will create a photo journal that represents significant features of the profession. Students may use any number of resources (but at least two different ones), including human resources, to obtain information about a science- or technology-related profession of interest to them. At a minimum, their research should include (1) a clear description of the profession and the work of the professional, (2) the amount of education or training needed for an entry-level position in the profession, and (3) the types of knowledge, specific skills and abilities, and attitudes and values needed to be successful in the profession.

Materials

- Library materials, brochures and career-related information from professional organizations, human resources, computers and software or Internet access, and similar resources

Presenting the Activity

Ask students how the knowledge they learn and the skills and dispositions they develop through their science classes might carry over to careers in these fields. (Some of these ideas may have already been discussed throughout the year, but now students will have an opportunity to investigate careers in science and technology in detail.)

Have students generate a list of science- and technology-related professions with which they are familiar. These may include the professions of a parent, relative, or neighbor. The list may include careers such as astronaut, architect, computer scientist, engineer, veterinarian, paleontologist, science teacher or professor, doctor, nurse, medical technician, pharmacist, or landscaper. Some students may identify careers related to telecommunications, health and nutrition, energy, aerodynamics, marine biology, plants and animals, environmental science, or other areas of science and technology they have studied.

Introduce the research project to students. Allow them to select a profession related to an area of science or technology they studied during the year (or to other areas of science, if you wish). They should develop a list of questions to research such as the following:

What is the profession called and what does the professional do?

What type and amount of education or training are necessary for the career?

What sort of knowledge base does the professional need?

What skills are required for this profession? Are there any certifications or licenses required?

What kinds of tools and technology are used in the work?

What is society's attitude toward this profession?

What attitudes or values would one need to be successful in this profession?

What is the value of the profession to society?

Students should record the information they find. Then, they should select four significant features of the profession for their photo journal. Each of these features should be represented by a "photograph" (drawing) of some aspect of the profession or of themselves and/or other professionals involved in working in the profession. Students should draw each picture in one of the "cameras" provided and describe the significant feature of the "photo." The four pictures and descriptions will represent a photo journal of significant features of the profession. In Activity 2, students will share their photo journals and their findings with others.

ACTIVITY 1   Name: _____

                             Date: _____

A PHOTO JOURNAL OF MY PROFESSION

Photo I

The profession I am presenting is _____

The science topic most closely related to it is _____

Title of entry: _____

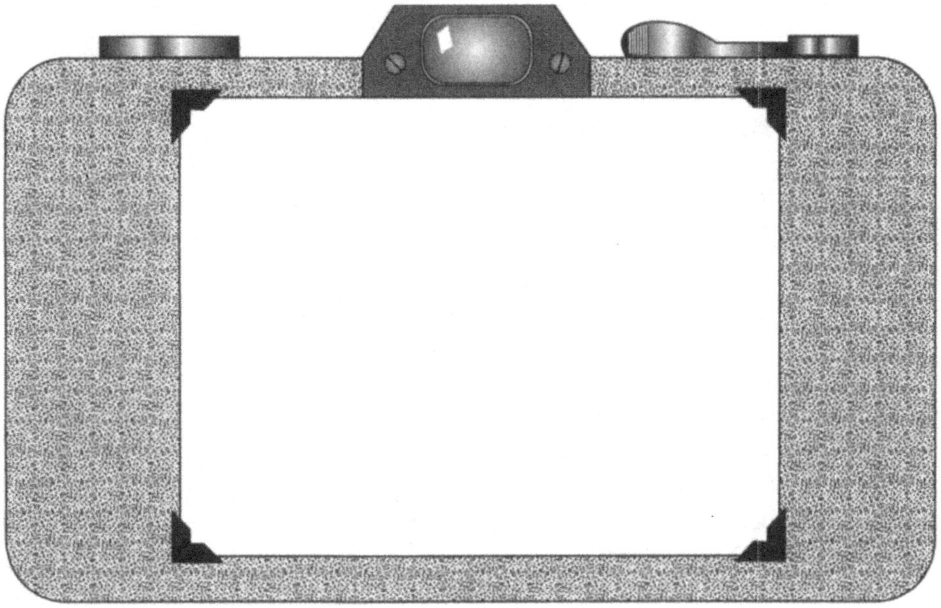

Description of photo: _____

_____

_____

_____

_____

_____

_____

_____

_____

_____

Name: _____

Date: _____

A PHOTO JOURNAL OF MY PROFESSION

Photo II

The profession I am presenting is _____

The science topic most closely related to it is _____

Title of entry: _____

Description of photo: _____

_____

_____

_____

_____

_____

_____

_____

_____

_____

Name: _____

Date: _____

A PHOTO JOURNAL OF MY PROFESSION

Photo III

The profession I am presenting is _____

The science topic most closely related to it is _____

Title of entry: _____

Description of photo: _____

_____

_____

_____

_____

_____

_____

_____

_____

Name: _____

Date: _____

A PHOTO JOURNAL OF MY PROFESSION

Photo IV

The profession I am presenting is _____

The science topic most closely related to it is _____

Title of entry: _____

Description of photo: _____

_____

_____

_____

_____

_____

_____

_____

_____

DESIGNING AND PERFORMING A CREATIVE INTERVIEW

Description of Activity

In this activity, students will share the significant features of the science- or technology-related career they researched by interviewing for a position in a most creative way. Instead of someone asking questions of them, they will give a presentation showing what they know and describing how they are qualified for the position. They may use their photo journals, create a brochure or poster, design a portfolio of resources and drawings, plan a multimedia presentation, compose a song or rap, choreograph a dance, or design another creative way to interview for a job in the profession they have selected to research.

Materials

- AV equipment (camcorder, VCR and monitor, audiotape or CD player, overhead projector, computer with PowerPoint); musical instruments (as needed)
- Audiotapes, videotapes, transparencies, CDs, and so forth (as needed)
- Poster board, drawing paper, art supplies
- Colored markers
- Other materials and props for presentations

Presenting the Activity

Tell students that they are going to interview for a position in the profession they researched. Since the competition for professional positions in most of the science and technology fields is high, they will need to develop and present a unique and exciting way to convince future employers that they have the knowledge, skills, and disposition needed for the position they are seeking. Explain that photo journals, creative portfolios, multimedia presentations, songs or raps, a dance, or other props may be included in their presentations. Their presentations should include evidence that they have the following

- The necessary knowledge base; technical and thinking skills; and attitudes, values, and other habits of mind needed to be successful in the profession
- An understanding of the relationship of the profession to science and technology
- An awareness of the importance of the profession to society

There are no student response pages for this activity. The product of this activity will be in the form of a portfolio, brochure, poster, or other set of visuals with an explanation; a multimedia, PowerPoint, or other slide presentation; a performance (song, rap, dance, play), which may include visuals or other props that convey important information; or another creative approach. It is important that the presentation be convincing and include the necessary information, as it is an interview for a job.

ANALYTIC RUBRIC: PICTURE PERFECT PROFESSIONS

Indicators of Learning Relate to These Unifying Concepts and Processes:

- Systems, Order, and Organization
- Evidence, Models, and Explanations
- Constancy, Change, and Measurement
- Evolution and Equilibrium
- Form and Function
- Mathematics, Literacy, and Thinking Skills

(Items in parentheses identify some of the concepts and skills assessed through the task.)

SCALE

	Complete	Almost	Not Yet	Comments
Activity 1: A Photo Journal of My Profession **The student:** • Completed a report about the profession (Evidence, Models, and Explanation) • Made a prediction based on reasoning • Used a minimum of two sources of information • Included information relating to the research questions identified by the class • For each photo journal entry: Provided the name of the profession, area of science, and title for significant feature (S-T-S Relationships; Concept Understanding) • Drew a picture of each of four significant features of the profession (Concept Representation; Communication) • Gave a complete description of each of four photo journal entries (Concept Application; Descriptive Writing)				
Activity 2: Designing and Performing a Creative Interview **The student:** • Demonstrated an understanding of the knowledge base, skills, and dispositions needed for success in a chosen profession related to science or technology (Concept Understanding and Application)				

	Complete	Almost	Not Yet	Comments
• Identified the relationship of the profession to science and/or technology (Concept Understanding and Application; Relationship of Science to Technology and Society) • Identified the importance of the profession to society (Concept Understanding and Application; Relationship of Science to Technology and Society)				
Suggested Criteria for Presentation Teachers may wish to add criteria for the presentations dealing with organization, quality of messages, creativity, audience appeal, or nature of presentation (convincing, complete, and so forth) Students may provide feedback to one another related to these criteria as "critical friends."				

Scoring Scale	
Complete	Student exhibited the indicator
Almost	Student showed some evidence that the indicator was exhibited, but something is incorrect or missing
Not Yet	Student did not show evidence of learning for that indicator

Appendix

Answers to Criterion-Referenced Tests

Chapter 5

1. c
2. c
3. a
4. d
5. c
6. a
7. a
8. a
9. b

Chapter 6

1. a
2. b
3. a
4. c
5. a
6. c

Chapter 7

1. b
2. a
3. c
4. d
5. c
6. a
7. a
8. c
9. a

Chapter 8

1. c
2. a
3. d
4. a
5. b
6. a
7. c
8. a
9. c
10. a
11. c
12. c
13. a

Chapter 9

1. a
2. b
3. d
4. a
5. c
6. a
7. a
8. b

Chapter 11

1. a
2. b
3. d
4. a
5. c
6. a
7. d
8. a
9. b

References and Bibliography

American Association for the Advancement of Science. (1989). *Science for all Americans*. New York: Oxford University Press.

American Association for the Advancement of Science. (1993). *Benchmarks for science literacy*. New York: Oxford University Press.

American Council on the Teaching of Foreign Languages. (1982). *ACTFL provisional proficiency guidelines*. Hastings-on-Hudson, NY: ACTFL Materials Center.

American Psychological Association. (1985). *Standards for educational and psychological testing*. Washington, DC: Author.

Anderson, S. R. (1993, June). Trouble with testing. *The American School Board Journal, 180*, 24–26.

Astin, A. W. (1991). *Assessment for excellence: The philosophy and practice of assessment and evaluation in higher education*. New York: American Council on Education/ Macmillan.

Baker, E. L. (1993, December). Questioning the technical quality of performance assessment. *The School Administrator, 50*, 12–16.

Baker, E. L., Aschbacher, P. R., Niemi, D., Yamaguchi, E., & Ni, Y. (1991). *Cognitively sensitive assessments of student writing in the content areas*. Los Angeles: The National Center for Research on Evaluation, Standards, and Student Testing (CRESST).

Baker, E. L., O'Neil, H. F., Jr., &. Linn, R. L. (1993). Policy and validity prospects for performance-based assessment. *American Psychologist, 48*(12), 1210–1218.

Berlak, H., Newmann, F. M., Adams, E., Archbald, D. A., Burgess, T., Raven, J., et al. (1992). *Toward a new science of educational testing and assessment*. New York: State University of New York Press.

Black, P., & Wiliam, D. (1998). Inside the black box: Raising standards through classroom assessment. *Phi Delta Kappan, 79*(2), 139–148.

Bloom, B. S. (Ed.). (1956). *Taxonomy of educational objectives. Book 1: Cognitive domain*. New York: Longman.

Bloom, B. S., Madaus, G. F., & Hastings, J. T. (1981). *Evaluation to improve learning*. New York: McGraw-Hill.

Bond, L., Friedman, L., & van der Ploeg, A. (1993). *Surveying the landscape of state educational assessment programs*. Washington, DC: Council for Educational Development and Research and the National Education Association.

Bracey, G. W. (1993, December). Testing the tests. *The School Administrator, 50*, 8–11.

Brandt, R. (Ed.). (1992). *Performance assessment: Readings from educational leadership*. Alexandria, VA: Association for Supervision and Curriculum Development.

Burke, K. (2005). *How to assess authentic learning* (4th ed.). Thousand Oaks, CA: Corwin Press.

Caine, R. N., & Caine, G. (1991). *Teaching and the human brain*. Alexandria, VA: Association for Supervision and Curriculum Development.

Caine, R. N., & Caine, G. (1997). *Education on the edge of possibility.* Alexandria, VA: Association for Supervision and Curriculum Development.

California Assessment Program. (1989). *Guidelines for the mathematics portfolio: Phase II pilot working paper.* Sacramento: California Assessment Program (CAP) Office, California State Department of Education.

California State Department of Education. (1989a). *A question of thinking: A first look at students' performance on open-ended questions in mathematics.* Sacramento, CA: Author.

California State Department of Education. (1989b). *Writing achievement of California eighth graders: Year two.* Sacramento, CA: Author.

Center on Learning, Assessment, and School Structure. (1993). *Standards, not standardization. Vol. III: Rethinking student assessment.* Geneseo, NY: Author.

Charles, R., & Silver, E. (Eds.). (1988). *The teaching and assessing of mathematical problem solving* (Vol. 3). Reston, VA: National Council of Teachers of Mathematics.

Committee on Education. (2004). *Lost in space: Science education in New York City public schools.* Report from the Council of the City of New York. New York: Author.

Connecticut Department of Education. (1990). *Toward a new generation of student outcome measures: Connecticut's common core of learning assessment.* Hartford, CT: Author.

Cox, D. (1987). *Concept-based science instruction.* Presentation, Portland State University, Portland, Oregon.

Cronbach, L. J. (1989). *Essentials of psychological testing* (5th ed.). New York: Harper & Row.

Department of Education and Science and the Welsh Office (UK). (1988). *National curriculum: Task group on assessment and testing: A report.* London: Author.

Department of Education and Science and the Welsh Office. (1989). *English for ages 5 to 16: Proposals of the secretary of state for education and science.* London: Author.

Educational Testing Service. (1986). *The redesign of testing for the 21st century: Proceedings of the ETS 1985 invitational conference.* Princeton, NJ: Author.

Educational Testing Service. (1993a). *Linking assessment with reform: Technologies that support conversations about student work.* Princeton, NJ: Author.

Educational Testing Service. (1993b). *What we can learn from performance assessment for the professions: Proceedings of the ETS 1992 invitational conference.* Princeton, NJ: Author.

Elbow, P. (1981). *Writing with power: Techniques for mastering the writing process.* New York: Oxford University Press.

Elbow, P. (1986). *Embracing contraries: Explorations in learning and teaching.* New York: Oxford University Press.

Elwell, P. (1991). To capture the ineffable: New forms of assessment in higher education. In G. Grand (Ed.), *Review of research in education.* Washington, DC: American Educational Research Association.

Enger, S., & Yager, R. E. (2001). *Assessing student understanding in science.* Thousand Oaks, CA: Corwin Press.

Falk, B., & Darling-Hammond, L. (1993). *The primary language record at P. S. 261: How assessment transforms teaching and learning.* New York: NCREST, Columbia University.

Feuer, M. J., Fulton, K., & Morrison, P. (1993, March). Better tests and testing practices: Options for policy makers. *Phi Delta Kappan, 74,* 532.

Finch, F. L. (Ed.). (1991). *Educational performance assessment.* Chicago: Riverside/Houghton Mifflin.

Fox, R. F. (1993, December). Do our assessments pass the test? *The School Administrator, 50,* 6–11.

Frederiksen, J. R., & Collins, A. (1989, December). A systems approach to educational testing. *Educational Researcher, 18*, 27–32.

Gardner, H. (1989). Assessment in context: The alternative to standardized testing. In B. Gifford (Ed.), *Changing assessments: Alternative views of aptitude, achievement, and instruction.* Boston: Kluwer Academic.

Gardner, H. (1993). *Multiple intelligences: The theory in practice.* New York: Basic Books.

Gardner, H. (1999). *Intelligence reframed: Multiple intelligences for the 21st century.* New York: Basic Books.

Gentile, C. (1991). *Exploring new methods for collecting students' school-based writing.* Washington, DC: U.S. Department of Education.

Gess-Newsome, J., & Lederman, N. (1999). *Examining pedagogical content knowledge.* Boston: Kluwer Academic.

Glaser, R. (1971). A criterion-referenced test. In J. Popham (Ed.), *Criterion-referenced measurement: An introduction.* Englewood Cliffs, NJ: Educational Technology.

Grant, G. (1979). *On competence: A critical analysis of competence-based reforms in higher education.* San Francisco: Jossey-Bass.

Gregory, G., & Parry, T. (2006). Designing brain-compatiable learning (3rd ed.). Thousand Oaks, CA: Corwin Press.

Hammerman, E. (2005a). *Eight essentials of inquiry-based learning.* Thousand Oaks, CA: Corwin Press.

Hammerman, E. (2005b). Linking classroom instruction and assessment to standardized testing. *Science Scope, 28*(4), 26–32.

Hammerman, E. (2006a). *Becoming a better science teacher.* Thousand Oaks, CA: Corwin Press.

Hammerman, E. (2006b). Toolkit for improving practice. *Science Scope, 30*(1), 18–23.

Haney, W. (1985, October). Making testing more educational. *Educational Leadership, 43*, 2.

Haney, W., & Scott, L. (1987). Talking with children about tests: An exploratory study of test item ambiguity. In K. O. Freedle & R. P. Duran (Eds.), *Cognitive and linguistic analyses of test performance.* Norwood, NJ: Ablex.

Hanson, F. A. (1993). *Testing testing: Social consequences of the examined life.* Los Angeles: University of California Press.

Hart, D. (1994). *Authentic assessment: A handbook for educators.* Menlo Park, CA: Addison-Wesley.

Herman, J., Aschbacher, P., & Winters, L. (1992). *A practical guide to alternative assessment.* Alexandria, VA: Association for Supervision and Curriculum Development.

International Baccalaureate Examination Office. (1991). *Extended essay guidelines.* Wales, UK: Author.

International Reading Association & the National Council of Teachers of English. (1996). *Standards for the English language arts.* Urbana, IL: NCTE.

Jorgensen, M. (1993, December). The promise of alternative assessment. *The School Administrator, 50*, 17–23.

Kendall, J. S., & Marzano, R. J. (1994). *The systematic identification and articulation of content standards and benchmarks* (Update). Aurora, CO: Mid-Continent Regional Educational Laboratory.

Kentucky General Assembly. (1990). *Kentucky Education Reform Act (KERA).* House Bill 940.

Lamme, L. L., & Hysmith, C. (1991, December). One school's adventure into portfolio assessment. *Language Arts, 68*, 629–640.

Lantz, Jr., H. B. (2004). *Rubrics for assessing student achievement in science Grades K–12.* Thousand Oaks, CA: Corwin Press.

Linn, R. (1986). Barriers to new test designs. In *The redesign of testing for the 21st century: Proceedings of the 1985 ETS invitational conference.* Princeton, NJ: Educational Testing Service.

Linn, R. (1989). *Educational measurement* (3rd ed.). Washington, DC: American Council on Education.

Linn, R. (1993). Educational assessment: Expanded expectations and challenges. *Educational Evaluation and Policy Analysis, 15*(1) 1–16.

Linn, R., Baker, E., & Dunbar, S. (1991, November). Complex, performance-based assessment: Expectations and validation criteria. *Educational Researcher, 20,* 15–21.

Lowell, A. L. (1926, January). The art of examination. *Atlantic Monthly, 137,* 58–66.

Maeroff, G. (1991, December). Assessing alternative assessment. *Phi Delta Kappan, 17,* 272–281.

Marzano, R. J. (1991). Fostering thinking across the curriculum through knowledge restructuring. *Journal of Reading, 34*(7), 518–525.

Marzano, R. J., Pikering, D. J., & McTighe, J. (1994). *Assessing student outcomes: Performance assessment using the dimensions of learning model.* Alexandria, VA: Association for Supervision and Curriculum Development.

Marzano, R. J., Pickering, D. J., & Pollock, J. E. (2001). *Classroom instruction that works.* Alexandria, VA: Association for Supervision and Curriculum Development.

Mehrens, W. (1992, Spring). Using performance assessment for accountability purposes. *Educational Measurement: Issues and Practice, 11,* 3–9.

Messick, S. (1989a). Meaning and values in test validation: The science and ethics of assessment. *Educational Researcher, 18*(2), 5.

Messick, S. (1989b). Validity. In R. Linn (Ed.), *Educational measurement* (3rd ed., pp. 13–103). Washington, DC: American Council on Education.

Mills, R. (1989, December). Portfolios capture rich array of student performance. *The School Administrator, 11,* 46.

Ministry of Education, Victoria, Australia. (1990). *Literacy profiles handbook: Assessing and reporting literacy development.* Victoria, Australia: Education Shop; distributed in United States by TASA, Brewster, NY.

Ministry of Education, Victoria, Australia. (1991). *English profiles handbook: Assessing and reporting students' progress in English.* Victoria, Australia: Education Shop; distributed in United States by TASA, Brewster, NY.

Mitchell, R. (1992). *Testing for learning.* New York: Free Press/Macmillan.

Moss, P. A. (1994, March). Can there be validity without reliability? *Educational Researcher, 23,* 5–12.

Musial, D. (2002). From where have all the standards come? *Thresholds in Education, 28*(4), 3–9.

Musial, D. (1996). Authentic assessment in problem-based learning. *Problem-Based Learning* Journal, *1*(2), 3–5.

Musial, D., & Hammerman, E. (1993). Knowledge through moments: A model for teaching thinking in science. *Teaching Thinking and Problem Solving, 14*(2) 12–15.

National Assessment of Educational Progress. (1987). *Learning by doing: A manual for teaching and assessing higher-order thinking in science and mathematics.* Princeton, NJ: Educational Testing Service.

National Center for Research on Evaluation, Standards, and Student Testing (CRESST). (1993, Winter). *The CRESST Line.*

National Center for Restructuring Education, Schools, and Teaching. (1993). *Authentic assessment in practice: A collection of portfolios, performance tasks, exhibitions, and documentations.* New York: Teachers College Press.

National Commission on Testing and Public Policy (NCTPP). (1990). *From gatekeeper to gateway: Transforming testing in America.* Chestnut Hill, MA: Author.

National Research Council. (1996). *National science education standards.* Washington, DC: National Academy Press.

National Research Council. (2000). *How people learn,* Washington, DC: National Academy Press.

National Research Council, Mathematical Sciences Education Board. (1993). *Measuring up: Prototypes for mathematics assessment.* Washington, DC: National Academy Press.

National Science Resources Center. (1997). *Science for all.* Washington, DC: National Academy Press.

NCTM Commission on Teaching Standards for School Mathematics. (1991). *Professional standards for teaching mathematics.* Reston, VA: National Council of Teachers of Mathematics.

NCTM Standards 2000 Project Writing Group. (2000). Principles and standards for school mathematics. Reston, VA: National Council of Teachers of Mathematics.

Perrone, V. (Ed.). (1991). *Expanding student assessment for supervision and curriculum development.* Alexandria, VA: Association for Supervision and Curriculum Development.

Purves, A. C. (1993). Setting standards in the language arts and literature classroom and the implications for portfolio assessment. *Educational Assessment, 1*(3), 175–199.

Redding, N. (1992, May). Assessing the big outcomes. *Educational Leadership, 49,* 49–51, 53.

Resnick, D. P., & Resnick, L. B. (1985, April). Standards, curriculum, and performance: A historical and comparative perspective. *Educational Researcher, 14,* 5–21.

Resnick, L. B. (1987). *Education and learning to think.* Washington, DC: National Academy Press.

Resnick, L. B., & Resnick, D. P. (1991). Assessing the thinking curriculum: New tools for educational reform. In B. Gifford (Ed.), *Changing assessments: Alternative views of aptitude, achievement and instruction.* Boston: Kluwer Academic.

Rogers, G. (1988). *Validating college outcomes with institutionally developed instruments: Issues in maximizing contextual validity.* Milwaukee, WI: Office of Research and Evaluation, Alverno College.

Rutherford, F. J., & Algren, A (1990). *Science for all Americans.* New York: Oxford University Press.

Sadler, D. R. (1987). Specifying and promulgating achievement standards. *Oxford Review of Education, 13*(2), 191–209.

Schwartz, J. L., & Viator, K. A (Eds.). (1990). *The prices of secrecy: The social, intellectual, and psychological costs of testing in America. A report to the Ford Foundation.* Cambridge, MA: Educational Testing Center, Harvard Graduate School of Education.

Shavelson, R. S., & Baxter, G. P. (1992, May). What we've learned about assessing hands-on science. *Educational Leadership, 49,* 20–25.

Shepard, L. (1989, April). Why we need better assessments. *Educational Leadership, 46,* 4–9.

Shoenfeld, A. H. (1988). Problem solving in context(s). In R. Charles & E. Silver (Eds.), *The teaching and assessing of mathematical problem solving.* Reston, VA: National Council of Teachers of Mathematics.

Sizer, T. R. (1984). *Horace's compromise: The dilemma of the American high school.* Boston: Houghton Mifflin.

Sizer, T. R. (1992). *Horace's school: Redesigning the American high school.* Boston: Houghton Mifflin.

Smith, C. B. (1993, December). Assessing job readiness through portfolios. *The School Administrator, 50,* 26–31.

Speech Communication Association. (1982). *Speaking and listening competencies for high school graduates.* Annandale, VA: Speech Communication Association of America.

Stiggins, R. (1991, March). Assessment literacy. *Phi Delta Kappan, 71,* 534–539.

Stiggins, R. (1994). *Student-centered classroom assessment.* New York: Macmillan.

Terenzini, P. (1987). The case for unobtrusive measures. In *Assessing the outcomes of higher education: Proceedings of the 1986 ETS invitational conference.* Princeton, NJ: Educational Testing Service.

Tierney, R., Carter, M., & Desai, L. (1991). *Portfolio assessment in the reading-writing classroom.* Norwood, MA: Christopher-Gordon.

University of California. (1989). *Assessment alternatives in mathematics.* From EQUALS and the California Mathematics Council, Lawrence Hall of Science. Berkeley, CA: Author.

U.S. Congress. (1992). Office of Technology Assessment. *Testing in American schools: Asking the right questions.* OTA SET-519. Washington, DC: Government Printing Office.

U.S. Department of Labor. (1991). *What work requires of schools: A SCANS (Secretary's Commission on Achieving Necessary Skills) report for America 2000.* Washington, DC: Government Printing Office.

Weiss, R. (2003). *Looking inside the classroom: A study of K–12 mathematics and science education in the United States.* Chapel Hill, NC: Horizon Research. Available at http://www.horizon-research.com/insidetheclassroom/

Wiggins, G. (1988, Winter). Rational numbers: Toward grading and scoring that help rather than harm learning. *American Educator, 12,* 20–48.

Wiggins, G. (1989a, May). A true test: Toward more authentic and equitable assessment. *Phi Delta Kappan, 70,* 703–713.

Wiggins, G. (1989b). Teaching to the (authentic) test. *Educational Leadership, 46* (April): 41–47.

Wiggins, G. (1991, February) Standards, not standardization: Evoking quality student work. *Educational Leadership, 48,* 18–25.

Wiggins, G. (1992, May). Creating tests worth taking. *Educational Leadership, 49,* 26–33.

Wiggins, G. (1993a). *Assessing student performance: Exploring the purpose and limits of testing.* San Francisco: Jossey-Bass.

Wiggins, G. (1993b, November). Assessment: Authenticity, context, and validity. *Phi Delta Kappan, 75,* 200–214.

Wiggins, G. (1994a, June). The immorality of test security. *Educational Policy, 8,* 157–182.

Wiggins, G. (1994b, July). None of the above. *The Executive Educator, 16,* 14–18.

Wiggins, G. (1998). *Educative assessment: Designing assessments to improve student performance.* San Francisco: Jossey-Bass.

Wittrock, M. C., & Baker, E. L. (1991). *Testing and cognition.* New York: Prentice Hall.

Wolf, D. (1988, December/January). Opening up assessment. *Educational Leadership, 44,* 4.

Wolf, D. (1989, April). Portfolio assessment: Sampling student work. *Educational Leadership, 46,* 35–39.

Wolf, D., Bixby, J., Glen, J. III, & Gardner, H. (1991). To use their minds well: Investigating new forms of student assessment. In G. Grant (Ed.), *Review of research in education.* Washington, DC: American Educational Research Association.

Zacharias, J. R. (1979, January). The people should get what they need: An optimistic view of what tests can be. *National Elementary Principal, 58,* 48–51.

Note: The May 1992 and April 1989 issues of *Educational Leadership,* the February 1992 issue of *Social Education,* the May 1989 issue of *Phi Delta Kappan,* and the December 1989 issue of *Educational Researcher* are devoted to assessment and testing reform issues.

Index

CORWIN PRESS

The Corwin Press logo—a raven striding across an open book—represents the union of courage and learning. Corwin Press is committed to improving education for all learners by publishing books and other professional development resources for those serving the field of PreK–12 education. By providing practical, hands-on materials, Corwin Press continues to carry out the promise of its motto: **"Helping Educators Do Their Work Better."**

Printed in the United States
By Bookmasters